ADVANCED TOPICS ON CELLULAR SELF-ORGANIZING NETS AND CHAOTIC NONLINEAR DYNAMICS TO MODEL AND CONTROL COMPLEX SYSTEMS

WORLD SCIENTIFIC SERIES ON NONLINEAR SCIENCE

Editor: Leon O. Chua
University of California, Berkeley

Series A. MONOGRAPHS AND TREATISES

Volume 43: Universality and Emergent Computation in Cellular Neural Networks
R. Dogaru

Volume 44: Bifurcations and Chaos in Piecewise-Smooth Dynamical Systems
Z. T. Zhusubaliyev & E. Mosekilde

Volume 45: Bifurcation and Chaos in Nonsmooth Mechanical Systems
J. Awrejcewicz & C.-H. Lamarque

Volume 46: Synchronization of Mechanical Systems
H. Nijmeijer & A. Rodriguez-Angeles

Volume 47: Chaos, Bifurcations and Fractals Around Us
W. Szemplińska-Stupnicka

Volume 48: Bio-Inspired Emergent Control of Locomotion Systems
M. Frasca, P. Arena & L. Fortuna

Volume 49: Nonlinear and Parametric Phenomena
V. Damgov

Volume 50: Cellular Neural Networks, Multi-Scroll Chaos and Synchronization
M. E. Yalcin, J. A. K. Suykens & J. P. L. Vandewalle

Volume 51: Symmetry and Complexity
K. Mainzer

Volume 52: Applied Nonlinear Time Series Analysis
M. Small

Volume 53: Bifurcation Theory and Applications
T. Ma & S. Wang

Volume 54: Dynamics of Crowd-Minds
A. Adamatzky

Volume 55: Control of Homoclinic Chaos by Weak Periodic Perturbations
R. Chacón

Volume 56: Strange Nonchaotic Attractors
U. Feudel, S. Kuznetsov & A. Pikovsky

Volume 57: A Nonlinear Dynamics Perspective of Wolfram's New Kind of Science
L. O. Chua

Volume 58: New Methods for Chaotic Dynamics
N. A. Magnitskii & S. V. Sidorov

Volume 59: Equations of Phase-Locked Loops
J. Kudrewicz & S. Wasowicz

Volume 60: Smooth and Nonsmooth High Dimensional Chaos and the Melnikov-Type Methods
J. Awrejcewicz & M. M. Holicke

Volume 61: A Gallery of Chua Attractors (with CD-ROM)
E. Bilotta & P. Pantano

WORLD SCIENTIFIC SERIES ON NONLINEAR SCIENCE Series A Vol. 63

Series Editor: Leon O. Chua

ADVANCED TOPICS ON CELLULAR SELF-ORGANIZING NETS AND CHAOTIC NONLINEAR DYNAMICS TO MODEL AND CONTROL COMPLEX SYSTEMS

Edited by

Riccardo Caponetto
Luigi Fortuna
Mattia Frasca
University of Catania, Italy

World Scientific

NEW JERSEY · LONDON · SINGAPORE · BEIJING · SHANGHAI · HONG KONG · TAIPEI · CHENNAI

Published by

World Scientific Publishing Co. Pte. Ltd.
5 Toh Tuck Link, Singapore 596224
USA office: 27 Warren Street, Suite 401-402, Hackensack, NJ 07601
UK office: 57 Shelton Street, Covent Garden, London WC2H 9HE

British Library Cataloguing-in-Publication Data
A catalogue record for this book is available from the British Library.

ADVANCED TOPICS ON CELLULAR SELF-ORGANIZING NETS AND CHAOTIC NONLINEAR DYNAMICS TO MODEL AND CONTROL COMPLEX SYSTEMS
World Scientific Series on Nonlinear Science, Series A — Vol. 63

ISBN-13 978-981-281-404-3
ISBN-10 981-281-404-3

Printed by FuIsland Offset Printing (S) Pte Ltd, Singapore

Preface

This book aims to discuss some of the topics investigated during the three years for the research project founded by the Italian Ministero dell'Istruzione, dell'Università e della Ricerca (MIUR) under the FIRB project RBNE01CW3M. The respective research units present the final results obtained in the project. Furthermore, the book aims to introduce new perspectives on the research on complexity. Many of the reported topics summarize the lectures given at the PhD National School "Reti Autoorganizzanti, Controllo e Sincronizzazione in Fenomeni Emergenti da Dinamiche Non Lineari" held at Bertinoro in July 2005.

April 2008

Contributors

Pierpaolo Andriani
Durham University Business School, UK

Michele Basso
Dipartimento di Sistemi e Informatica
Università degli Studi di Firenze

Stefano Boccaletti[1]
Istituto Nazionale di Ottica Applicata, Firenze

Maide Bucolo
Dipartimento di Ingegneria Elettrica Elettronica e dei Sistemi
Università degli Studi di Catania

Riccardo Caponetto
Dipartimento di Ingegneria Elettrica Elettronica e dei Sistemi
Università degli Studi di Catania

Mario Chavez[2]
Istituto Nazionale di Ottica Applicata, Firenze

Francesca Conti
Dipartimento di Ingegneria Elettrica Elettronica e dei Sistemi
Università degli Studi di Catania

[1]S. Boccaletti is now with CNR-Istituto dei Sistemi Complessi, Firenze, Italy
[2]M. Chavez is now with the Cognitive Neurosciences and Brain Imaging Laboratory LENA-CNRS-UPR-640. Hôpital de La Salpêtrière, Paris, France

Claudia Coronnello
Dipartimento di Fisica e Tecnologie Relative, Università degli Studi di Palermo and INFM-CNR, Unità di Palermo

Luigi Fortuna
Dipartimento di Ingegneria Elettrica Elettronica e dei Sistemi
Università degli Studi di Catania

Mattia Frasca
Dipartimento di Ingegneria Elettrica Elettronica e dei Sistemi
Università degli Studi di Catania

Manuela La Rosa
STMicroelectronics, Catania

Fabrizio Lillo
Dipartimento di Fisica e Tecnologie Relative, Università degli Studi di Palermo and INFM-CNR, Unità di Palermo, and Santa Fe Institute, Santa Fe, USA

Rosario Nunzio Mantegna
Dipartimento di Fisica e Tecnologie Relative, Università degli Studi di Palermo and INFM-CNR, Unità di Palermo

Donatello Materassi
Dipartimento di Sistemi e Informatica
Università degli Studi di Firenze

Salvatore Miccichè
Dipartimento di Fisica e Tecnologie Relative, Università degli Studi di Palermo and INFM-CNR, Unità di Palermo

Giuseppina Passiante
Department of Innovation Engineering, University of Lecce

Carlo Piccardi
Dipartimento di Elettronica e Informazione
Politecnico di Milano

Sergio Rinaldi
Dipartimento di Elettronica e Informazione
Politecnico di Milano

Alessandro Rizzo
Dipartimento di Elettrotecnica ed Elettronica,
Politecnico di Bari

Michele Tumminello
Dipartimento di Fisica e Tecnologie Relative,
Università degli Studi di Palermo

List of People Involved in the FIRB Project

ALLEN Peter, University of Cranfield

ANDRIANI Pierpaolo, Business School - University of Durham

ARECCHI Fortunato Tito, Università degli Studi di Firenze

ARENA Paolo, Università degli Studi di Catania

BAGNI Gianni, Università degli Studi di Firenze

BASILE Adriano, Università degli Studi di Catania

BASSO Michele, Università degli Studi di Firenze

BOCCALETTI Stefano, Istituto Nazionale di Ottica

BONANNO Giovanni, Istituto Nazionale Fisica della Materia

CAPONETTO Riccardo, Università degli Studi di Catania

CASAGRANDI Renato, Politecnico di Milano

CHUA Leon, University of California Berkeley

DERCOLE Fabio, Politecnico di Milano

FAZZINO Stefano, Università degli Studi di Catania

FORTI Mauro, Università degli Studi di Siena

FORTUNA Luigi, Università degli Studi di Catania

FRASCA Mattia, Università degli Studi di Catania

GALLO Antonio, Università degli Studi di Catania

GENESIO Roberto, Università degli Studi di Firenze

GIACOMELLI Gianni, Istituto Nazionale di Ottica

GORAS Liviu, The "Gh. Asachi" Technical University of Iasi, Romania

GRAGNANI Alessandra, Politecnico di Milano

LILLO Fabrizio, Istituto Nazionale Fisica della Materia

LIPPI Gianluca, INLN

LIVI Roberto, Università degli Studi di Firenze

MANTEGNA Rosario Nunzio, Istituto Nazionale Fisica della Materia

MEUCCI Riccardo, Istituto Nazionale di Ottica

NUNNARI Giuseppe, Università degli Studi di Catania

OCCHIPINTI Luigi Giuseppe, STMicroelectronics

PASSIANTE Giuseppina, Università degli Studi di Lecce

PICCARDI Carlo, Politecnico di Milano

PIKOVSKI Arkady, University of Potsdam

POLITI Antonio, Istituto Nazionale di Ottica

RAMAZZA Pierluigi, Istituto Nazionale di Ottica

RINALDI Sergio, Politecnico di Milano

ROSKA Tamas, HAS (Hungarian Accademy of Scienece)

RUFFO Stefano, Università degli Studi di Firenze

TESI Alberto, Università degli Studi di Firenze

TORCINI Alessandro, Università degli Studi di Firenze

ZAPPA Giovanni, Università degli Studi di Firenze

Contents

Chapter 1

The CNN Paradigm for Complexity

M. Bucolo, R. Caponetto, L. Fortuna and M. Frasca

Complex systems made of nonlinear interacting cells are gaining ever and ever interest. In this Chapter, a new generalized paradigmatic model based on Cellular Nonlinear Networks (CNN), for the study of these systems is presented. Starting from this model, an universal environment, E^3, for the study of such complex systems providing to the user the possibility of defining the dynamics of the cells as well as the connections among them in a simple and flexible way is proposed.

In this Chapter, the main characteristics of E^3 are discussed. This simulator is based on the paradigm of CNN and makes use of parallel computing allowing to considerably extend the size of the simulated system. Each cell is conceived as a nth-order dynamical system. Moreover, numerical results obtained by emulating a complex system made of a large number of interacting chaotic units arranged in a 3D lattice are presented. These simulations show an emerging result: organized forms and structure arise from the dynamic evolution of such systems. Furthermore, the Chapter stresses the universal role that CNNs are assuming today. It is shown that the dynamical behavior of 3D CNN-based models allows us to approach new emerging problems, to open new research frontiers as the generation of new geometrical forms and to establish some links between art, neuroscience and dynamical systems.

1.1 Introduction

In recent years complex systems are gaining ever and ever interest. This interest is witnessed by the large number of papers in which the term complexity is used. This term is often used in very different contexts (biology, architecture, computer science, engineering, physics) and has different meanings: it has been associated with the difficulty of description, the difficulty of creation, the degree of organization of a system [Lloyd (2001)]. Several scientists have analyzed complex systems and formulated very important concepts such as order from disorder [Schrodinger (1948)], self-organization [Nicolis (1997)], synergetics [Haken (1983)], cooperative phenomenon [Haken (1975)], edge of chaos [Langton (1990)]. These concepts are now key points of the theory of complexity. Moreover, recent developments have lead to new analyses based on quantitative tests [Chua (1998); Yang (2001)]. In all these studies the importance of numerical simulations to study complex systems is evident.

On the other hand, the study of complex systems lead to important, yet unanswered questions. Shapes are the fundamental attribute of visible objects; they give us the perception of the structure of them, and they also represent the alphabet of art [Arnheim (1974)]. Shapes represent the object of visible beauty. It is well known that the mathematics of shape and space is geometry. Which is the relationship between dynamical systems and shapes? Could the mathematics of nonlinear dynamics help us to establish an innovative way to go in depth into the world of shape generation?

The human mind process of creating shapes is of course a dynamical process and often its complexity involves more steps in combining emergent conditions. The shape generation in the human mind leads often to the emergence status of mental states and the brain-mind complexity is related also to its shape organization process.

Some artistic experiences and painters of the last century are taken as examples of the previous concepts: Joan Miró, Salvator Dalí art and Giorgio De Chirico. Both the artist's evolution trends and the process of shapes creation are emergent phenomena . When rich abstract shapes did appear in visual arts?

The generation of new shapes is due essentially to an emergent mechanism in the human mind. One of the tasks of this Chapter is to prove experimentally that shapes are also emergent phenomena Emergent phenomena in chaotic spatially extended systems. Moreover, emergent phenomena that occur in visual cortex could also be related to the previous items. In fact many researches on this subject focused on this topic [Ermentrout (1979); Dalhem (2000)]. The hallucination phenomena arising in illness like migraine lead to the generation of new forms known as phosphenes or scintillating scotoma, fortifications and vortices. New forms and structures emerge in the visual cortex without external inputs. Possibly these unpredictable events influenced positively some paints of famous artists like Giorgio De Chirico or Vincent Van Gogh [Podoll (2001); Bogousslavsky (2003)].

Furthermore, a very appealing approach to face the visual cortex behavior has been introduced in [Zeki (1999)], where the neuroscientist discusses the understanding of both the single visual cortex areas and its global organization, by analyzing various artists' paints. In this last remark the twofold link between art and neuroscience appears. In the actual literature these aspects are reinforced by recent studies on surprising emergent creative tendencies shown in patients with strong mental diseases [Giles (2004)]. Shapes are the links among these important routes.

In this Chapter, the characteristics of E^3, an universal environment for the emulation of complex systems, based on a new generalized CNN model, are discussed. Then, shapes generated by many examples of complex systems made of nonlinear interacting units are analyzed in order to show that the CNN paradigm [Chua (1998)] is an appropriate tool to generate emergent shape patterns , to relate shape evolution with spatially extended dynamical systems and therefore to open a real bridge between circuits, art and neuroscience.

The essential prerogative of the CNN paradigm is to take advantage of the cooperative behavior of simple dynamical nonlinear circuits in order to obtain complex global tasks. The reported study is related to three dimensional, in the space, CNN (3D-CNN). Merging cells to achieve E-merging patterns and dynamics and to emulate complex system behaviors fully expresses the 3D-CNN role in the proposed study. In fact from the coupling of the single 3D-CNN units impressive patterns and shapes emerge. It is shown that the evolution of the 3D-CNN dynamics generates harmonious yet unpredictable shapes. It is well known that a partial differential equation can be matched in a CNN-based algorithm [Chua (1998)] and that RD-CNNs are appropriate to reproduce complex phenomena in biology, chemistry, neurodynamics: the generalized 3D-CNN scheme proposed in this work adds further results in order to remark the power of the CNN paradigm.

1.2 The 3D-CNN Model

Let us consider the universal definition of a CNN [Chua (1998)]: *A CNN is any spatial arrangement of locally coupled cells where each cell is a dynami-*

cal system which has inputs, and outputs and a state evolving in accordance with the described dynamical laws.

The following general concepts will be assumed:

- The coupling laws are not generally spatially invariant (even if often for practical realizations these are assumed invariant).
- The concept of dominant local coupling is assumed; therefore most of the connections are between cells in a neighborhood of unitary radius, but some non-local connections could also be included.
- Each cell is a dynamical system with assigned state variables.

Let us consider an isolated cell (from the microscale CNN point of view). The following variables characterize the cell:

- the exogenous, controllable input vector $\mathbf{u}_{i,j,k}(t) \in \mathbb{R}^{m_u}$;
- the exogenous, uncontrollable input vector $\mathbf{S}_{i,j,k}(t) \in \mathbb{R}^{m_s}$;
- the state vector $\mathbf{x}_{i,j,k}(t) \in \mathbb{R}^{n}$;
- the output vector $\mathbf{y}_{i,j,k}(t) \in \mathbb{R}^{m_y}$;
- the bias vector $\mathbf{z}_{i,j,k}(t) \in \mathbb{R}^{n}$ that generally is assumed controllable;

where i, j, k represent the space variables.

Moreover, let us introduce some useful notations. Let us indicate with $\mathbf{X}$, $\mathbf{U}$, $\mathbf{Y}$ the whole state, input and output sets (referring to all the cells of the CNN). Let us define the neighborhood of the cell $C(i,j,k)$ as

$$N_r(i,j,k) = \{C(\alpha,\beta,\gamma)\, |\max(|\alpha - i|\,, |\beta - j|\,, |\gamma - k|) \leq r\}$$

and let us indicate with

$$\mathbf{X}_{i,j,k} = \bigcup_{C(\alpha,\beta,\gamma)\in N_r(i,j,k)} \mathbf{x}_{\alpha,\beta,\gamma}$$

the state variables of cells in the neighborhood $N_r(i,j,k)$ of the cell $C(i,j,k)$. Analogous definitions can be given for $\mathbf{U}_{i,j,k}$ and $\mathbf{Y}_{i,j,k}$.

The general CNN model that we introduce is built-up by adding complexity to the simplest CNN and takes into account the following key points:

- The cells do not require to be equal each other.
- In a 3D grid the coupling laws are locally described along with the neighbor cells $S_{\alpha,\beta,\gamma}$.
- Each node, as described below, can be realized using n generalized first order cells constituting a small multilayer CNN architecture.

More in detail, referring to the considerations above, we could give definitions of CNN structures at several levels, starting from the microscale level to the mesoscale and macroscale level.

Definition: microscale CNN

A CNN obtained by connecting first-order cells may implement any nonlinear dynamics [Fortuna (2003)]. The microscale CNN is introduced to account for this concept. It can be described by the following equations:

$$\begin{aligned}
\dot{x}_1 &= -x_1 + z_1 + \sum_{C(l)\in N_r(1)} (A(1;l)y_l + B(1;l)u_l + C(1;l)x_l) \\
\dot{x}_2 &= -x_2 + z_2 + \sum_{C(l)\in N_r(2)} (A(2;l)y_l + B(2;l)u_l + C(2;l)x_l) \\
&\dots \\
\dot{x}_n &= -x_n + z_n + \sum_{C(l)\in N_r(n)} (A(n;l)y_l + B(n;l)u_l + C(n;l)x_l) \\
y_1 &= 0.5(|x_1 + 1| - |x_1 - 1| \\
&\dots \\
y_n &= 0.5(|x_n + 1| - |x_n - 1|
\end{aligned} \tag{1.1}$$

For instance, a CNN made of three first order cells ($n = 3$) may implement the dynamics of the Chua's circuit [Arena (1995)] or of a Colpitts-like oscillator [Arena (1996)]. A schematic representation of a microscale CNN is shown in Fig. 1.1, where each first-order cell is represented by a small cube and the possibility of connecting such a CNN with other CNNs is sketched.

This model can be further generalized as follows:

$$\begin{aligned}
\dot{x}_{1,ijk} &= f_{1;i,j,k}(\mathbf{x}_{ijk}, \mathbf{y}_{ijk}, \mathbf{u}_{ijk}) \\
\dot{x}_{2,ijk} &= f_{2;i,j,k}(\mathbf{x}_{ijk}, \mathbf{y}_{ijk}, \mathbf{u}_{ijk}) \\
&\dots \\
\dot{x}_{n,ijk} &= f_{n;i,j,k}(\mathbf{x}_{ijk}, \mathbf{y}_{ijk}, \mathbf{u}_{ijk}) \\
y_{1,ijk} &= g_{1;i,j,k}(\mathbf{x}_{ijk}, \mathbf{y}_{ijk}, \mathbf{u}_{ijk}) \\
&\dots \\
y_{n,ijk} &= g_{n;i,j,k}(\mathbf{x}_{ijk}, \mathbf{y}_{ijk}, \mathbf{u}_{ijk})
\end{aligned} \tag{1.2}$$

where we make explicit the connections with other CNNs through the space variables i, j, k.

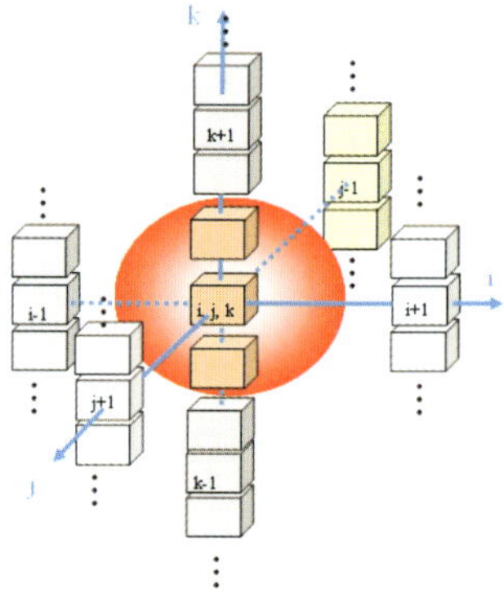

Fig. 1.1 Schematic representation of a microscale CNN.

Definition: mesoscale CNN

By connecting several microscale CNNs (1.2) in a 3D grid, a mesoscale CNN can be obtained. A microscale CNN is now a cell of a mesoscale CNN. The cell equations for a mesoscale CNN are defined as follows:

$$\begin{aligned} \dot{\mathbf{x}}_{ijk} &= f_{i,j,k}(\mathbf{x}_{ijk}, \mathbf{y}_{ijk}, \mathbf{u}_{ijk}) + a_{ijk}(\mathbf{Y}_{ijk}) + b_{ijk}(\mathbf{U}_{ijk}) + c_{ijk}(\mathbf{X}_{ijk}) \\ \mathbf{y}_{ijk} &= g_{i,j,k}(\mathbf{x}_{ijk}, \mathbf{y}_{ijk}, \mathbf{u}_{ijk}) \end{aligned} \tag{1.3}$$

where $a_{i,j,k}(\mathbf{Y}_{ijk})$, $b_{i,j,k}(\mathbf{U}_{ijk})$, $c_{i,j,k}(\mathbf{X}_{ijk})$ represent the coupling terms. It can be noticed that connections among cells are only local. A schematic view of a mesoscale CNN is shown in Fig. 1.2.

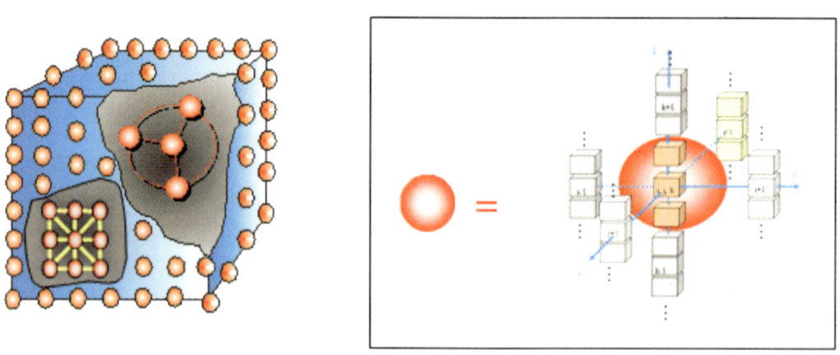

Fig. 1.2 Schematic representation of the generalized model of 3D-CNN.

Definition: macroscale CNN

We now add to this structure the possibility of having long-range connections. By including in Equation (1.3) these terms, the macroscale model of a CNN architecture assumes the following form:

$$\begin{aligned}\dot{\mathbf{x}}_{ijk} &= f_{i,j,k}(\mathbf{x}_{ijk}, \mathbf{y}_{ijk}, \mathbf{u}_{ijk}) + a_{i,j,k}(\mathbf{Y}_{ijk}) + b_{i,j,k}(\mathbf{U}_{ijk}) \\ &\quad + c_{i,j,k}(\mathbf{X}_{ijk}) + a^{lr}_{i,j,k}(\mathbf{Y}) + b^{lr}_{i,j,k}(\mathbf{U}) + c^{lr}_{i,j,k}(\mathbf{X}) \qquad (1.4) \\ \mathbf{y}_{ijk} &= g_{i,j,k}(\mathbf{x}_{ijk}, \mathbf{y}_{ijk}, \mathbf{u}_{ijk})\end{aligned}$$

where $a_{i,j,k}(\mathbf{Y}_{ijk})$, $b_{i,j,k}(\mathbf{U}_{ijk})$, $c_{i,j,k}(\mathbf{X}_{ijk})$ represent local coupling terms, while $a^{lr}_{i,j,k}(\mathbf{Y}_{ijk})$, $b^{lr}_{i,j,k}(\mathbf{U}_{ijk})$, $c^{lr}_{i,j,k}(\mathbf{X}_{ijk})$ take into account long-range coupling terms.

The overall model described by Equations (1.4) is schematized in Fig. 1.3. Figure 1.3 represents a schematic view of such a CNN, in which local and long-range connections are allowed. Moreover, the space dependency of each cell also includes the uncertainties in the model of each cell. These uncertainties can be either parametric or structural.

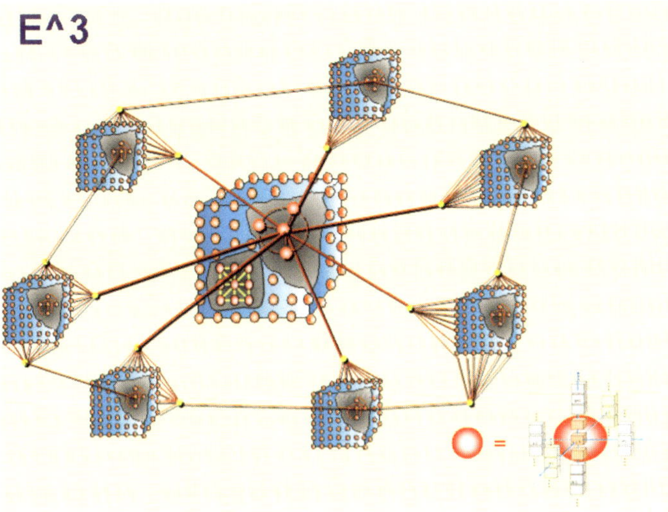

Fig. 1.3 Schematic representation of the generalized model of 3D-CNN.

This model can then be further generalized and represented by the following state equation:

$$\begin{aligned}\dot{\mathbf{x}}_{ijk} &= f_{i,j,k}(\mathbf{X},\mathbf{Y},\mathbf{U})\\ \mathbf{y}_{ijk} &= g_{i,j,k}(\mathbf{X},\mathbf{Y},\mathbf{U})\end{aligned} \tag{1.5}$$

The overall model (macroscale CNN) schematized in Figure 1.3 is built-up by adding complexity to the simplest CNN (the microscale CNN) through the mesoscale CNN in which connections are only local.

Example. 3D-CNN with first-order cells with standard nonlinearity

A macroscale 3D-CNN with first-order cells with standard nonlinearity is described by the following equations:

$$\begin{cases} \dot{x}_{ijk} = -x_{ijk} + z_{ijk} + \sum\limits_{C(\alpha,\beta,\gamma)\in N_r(i,j,k)} A(i,j,k;\alpha,\beta,\gamma)y_{\alpha,\beta,\gamma} \\ \qquad + \sum\limits_{C(\alpha,\beta,\gamma)\in N_r(i,j,k)} B(i,j,k;\alpha,\beta,\gamma)u_{\alpha,\beta,\gamma} \\ \qquad + \sum\limits_{C(\alpha,\beta,\gamma)\in N_r(i,j,k)} C(i,j,k;\alpha,\beta,\gamma)x_{\alpha,\beta,\gamma} \\ \qquad + A^{lr}_{i,j,k;\alpha,\beta,\gamma}(y_{\alpha,\beta,\gamma}) \\ \qquad + B^{lr}_{i,j,k;\alpha,\beta,\gamma}(u_{\alpha,\beta,\gamma}) \\ \qquad + C^{lr}_{i,j,k;\alpha,\beta,\gamma}(x_{\alpha,\beta,\gamma}) \\ y_{ijk} = 0.5(|x_{ijk}+1| - |x_{ijk}-1|) \end{cases} \tag{1.6}$$

where $A(i,j,k;\alpha,\beta,\gamma)$, $B(i,j,k;\alpha,\beta,\gamma)$ and $C(i,j,k;\alpha,\beta,\gamma)$ are the feedback template, the control template and the state template; $A^{lr}_{i,j,k;\alpha,\beta,\gamma}$, $B^{lr}_{i,j,k;\alpha,\beta,\gamma}$ and $C^{lr}_{i,j,k;\alpha,\beta,\gamma}$ describe the map of the long-range connections (they are NxNxN matrices). Equations (1.6) match equations (1.4) if the following assumptions on the coupling terms hold:

$$a^{lr}_{i,j,k}(\mathbf{Y}_{ijk}) = \sum_{C(\alpha,\beta,\gamma)\in N_r(i,j,k)} A^{lr}(i,j,k;\alpha,\beta,\gamma)y_{\alpha,\beta,\gamma}$$

$$a_{i,j,k}(\mathbf{Y}_{ijk}) = \sum_{C(\alpha,\beta,\gamma)\in N_r(i,j,k)} A(i,j,k;\alpha,\beta,\gamma)y_{\alpha,\beta,\gamma}$$

$$b^{lr}_{i,j,k}(\mathbf{Y}_{ijk}) = \sum_{C(\alpha,\beta,\gamma)\in N_r(i,j,k)} B^{lr}(i,j,k;\alpha,\beta,\gamma)y_{\alpha,\beta,\gamma}$$

$$b_{i,j,k}(\mathbf{Y}_{ijk}) = \sum_{C(\alpha,\beta,\gamma)\in N_r(i,j,k)} B(i,j,k;\alpha,\beta,\gamma) y_{\alpha,\beta,\gamma}$$

$$c^{lr}_{i,j,k}(\mathbf{Y}_{ijk}) = \sum_{C(\alpha,\beta,\gamma)\in N_r(i,j,k)} C^{lr}(i,j,k;\alpha,\beta,\gamma) y_{\alpha,\beta,\gamma}$$

$$c_{i,j,k}(\mathbf{Y}_{ijk}) = \sum_{C(\alpha,\beta,\gamma)\in N_r(i,j,k)} C(i,j,k;\alpha,\beta,\gamma) y_{\alpha,\beta,\gamma}$$

Remark. The conceived structure, showing emerging pattern behaviors by using cells, that means to start from simplicity to achieve complexity, is an appealing research subject in several fields. Let us consider the well known architect Eladio Dieste [Morales (1991)]; he used the brick as the fundamental element to build his main structures in civil engineering. The Chua' s cell [Chua (1988a,b); Fortuna (2003)] represents the basic electronic element to design complex circuits exactly like the Eladio Dieste's brick. Moreover as the architect Frei Otto [Otto (1982)] considered emerging forms in nature as building paradigms to be realized in civil structures, in a dual manner the search for complex patterns and their circuit realization by means of electronic structures find the link in the CNN architecture.

1.3 E^3: An Universal Emulator for Complex Systems

The model introduced in Section 1.2 was the basis for E^3, the framework developed for the study and the simulation of complex systems. In this Section the main features of E^3 are illustrated. This software has been designed to simulate complex systems made of interacting dynamic nonlinear units. Therefore, E^3 has been designed to include three properties: the dynamics of each simple unit is arbitrary, the cells of the complex system are either identical or different, connections are arbitrary.

E^3 provides the user with the possibility of implementing in a very simple way arbitrary dynamics for each cell of the complex systems. Following this approach each cell of the complex CNN consists of an arbitrary n-order system, and may be defined by the user by writing its equations. This differs from standard CNN in which each cell dynamics is defined by templates, which therefore contain both cell dynamics and connections between cells.

Moreover, the cells of the 3D-CNN may be different each other. The more general case is that the equations defining each of the cells constituting the complex systems are different from cell to cell. This is for example the case in which one would model fire propagation in two different adjacent substratums. Moreover, cells may be nearly identical. In this case cells of the complex systems may differ only for the value assumed by their characterizing parameters. This case is different from the previous case, in fact in this case one does not need to write new equations, but the possibility to have space-variant parameters should be included.

Finally, the connections among the units of the CNN and, in general, of a complex system may be of several types. Several examples of complex systems made of locally interacting units have been studied in literature; all-to-all coupling is also very common in modelling complex systems. While random networks efficiently model phenomena like stock markets and small world connections account for model such as spread of diseases, modelling the structure of the world wide web requires dynamically changing network (scale-free network). A general simulator for complex systems should provide the possibility of implementing all these structures in an easy way and at the same time should allow the user the possibility of reconnecting arbitrary cells of the system.

Another important characteristic of E^3 is the use of parallel computing. A general structure has been designed for E^3. The simulator can be run either on a single machine or on a network of personal computers. This second case implements parallel computing. The software has been written in C (routines for numerical integration, output visualization, definition of cell dynamics) and Java (communication between processors, management of data distribution) and is based on open-source components.

Parallel computing may be realized by using *coarse grained* or *fine grained* architectures. A coarse grained architecture refers to the case in which the computation is distributed to a small number of high-capability processors, while the architecture is fine grained when an high number of simple processors is used. In this sense a CNN is the most significative example of fine grained architecture. The architecture of E^3 is based on a small number of high-capability processors.

The strategy adopted to implement parallel computing is the so-called *domain decomposition*, in which data are distributed among the processors executing the same operations on different portions of the data. In fact the problem of emulating a system made of many units can be simply decomposed into several domains made of subparts of the whole set of the cells. A processor plays the role of master and collects all the results coming from the elaboration by the other processors. Moreover, through the so-

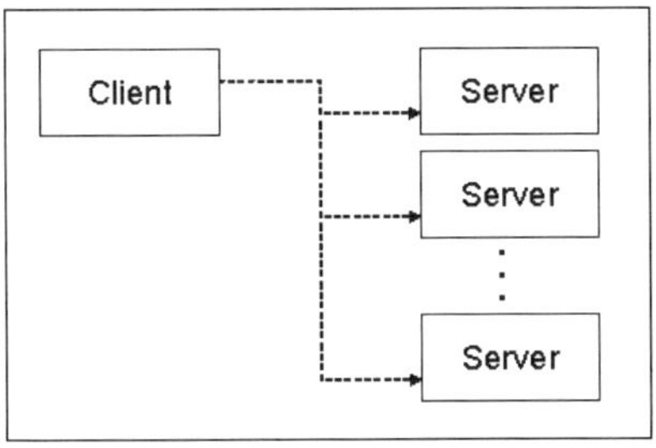

Fig. 1.4 Main modules of the software architecture of E^3.

called *message passing* each processor may obtain data processed by other units of the parallel architecture.

A very efficient way to implement message passing is to create a cluster of workstation in a LAN network. The whole software therefore consists of two main modules, called server and client (Figure 1.4). The server runs in each processor of the network, while the client runs on the master PC, coordinating the data coming from the different processors. The complex system to be emulated has to be defined in the client. The first operation executed by the client is to create the structure and to assign to each clients a portion of the cells to be simulated. Then the integration routine is performed.

To allow the definition of the cell dynamics by the user it has been chosen to implement a routine called *compile* able to create a .dll file starting from the cell equations written in C language. The integration routine uses the .dll file: it is created by the client and then distributed to each server. This choice has been adopted to create an universal tool that does not require long training to be used. Practically, the only code that the user has to write refers to the cell equations.

As regards the topology of connections it has been chosen to implement the more general case, i.e. to not restrict to the case of local connections, by allowing to the user the possibility of customizing each connection. The whole set of connections is described in the .net file, that can be modified by the user. Moreover, several routines to generate the most commonly used topologies (local connections in 1D, 2D and 3D spaces, all-to-all coupling, and so on) have been designed. These routines create the .net file that can be successively edited to introduce modifications in the simulated system.

E^3 provides several built-in procedures for output visualization such as 3D level surface and 2D density plots. Moreover, all the data can be stored in universal format for further evaluation.

The input files for the client module of E^3 are reassumed in the following:

- the .mpi file containing the parameters defining communication between client and server;
- the .net file containing the network topology;
- the .out file containing the parameters for the numerical integration and output parameters.

1.4 Emergence of Forms in 3D-CNNs

Looking at a picture and in general at an artistic representation, a great number of impressions are received, all at the same time. Let us consider a paint: a lot of elements appear, the impressions that are received combining the various elements give the perception of the specific characteristic of the paint. The unifying principle of the different elements is the shape that expresses the whole: the shape is an emergent property of the whole.

In the 3D-CNN paradigm the single contributions from the various cells lead us to the global characteristics of the architecture as regards both the structure and the emergent dynamical behavior. There is a dichotomy in this view between CNNs and shapes. The idea that the dynamics of a 3D-CNN is strongly evolving is assumed. Moreover, we cannot directly perceive the internal dynamical evolution of the single cell or of a group of cells that leads us to understand the whole emergency of the CNN behavior, that could only be synthetized into a shape. The internal global dynamics could be only perceived by coupling to the 3D-CNN the vision of the shapes. A dichotomy does exist: the shape generation capabilities of CNNs are emphasized, moreover the shape as partial abstraction of the emergent property in a distributed nonlinear dynamical system does appear. These concepts are the leading points of the next part of this Section where in details both the configurations and the visual representations of the generalized 3D-CNN paradigm are included.

The model used in the following experiments is related to a very simplified coupling law, where a simple space-constant diffusion law in Equations (1.2) is assumed as follows:

$$c_{i,j,k}(\mathbf{X}_{ijk}) = D\nabla^2_{ijk}\mathbf{x} \tag{1.7}$$

where the discretized Laplacian in a 3D space is defined by the following relationship:

$$\nabla^2_{ijk}\mathbf{x} = \mathbf{x}_{i-1,jk} + \mathbf{x}_{i+1,jk} + \mathbf{x}_{i,j-1,k} + \mathbf{x}_{i,j+1,k} + \mathbf{x}_{ij,k-1} + \mathbf{x}_{ij,k+1} - 6\mathbf{x}_{ijk}$$

Moreover, all the cells are equal, i.e. $f_{i,j,k}(\mathbf{x}_{ijk}, \mathbf{u}_{ijk}) = f(\mathbf{x}_{ijk}, \mathbf{u}_{ijk})$.

With these assumptions equations (1.3) can therefore be rewritten as follows:

$$\dot{\mathbf{x}}_{ijk} = f(\mathbf{x}_{ijk}) + D\nabla^2_{ijk}\mathbf{x}$$

which matches the well-known paradigm of reaction-diffusion equations:

$$\dot{\mathbf{x}} = f(\mathbf{x}) + D\nabla^2\mathbf{x}$$

The experiments discussed below have been performed considering different dynamical laws for each cell of the CNN. In all the cases the behavior of each cell is chaotic. The coupling law is characterized by low coupling coefficient (weak diffusion). In all the experiments zero-flux boundary conditions have been chosen.

The experiments will be detailed in the following. In all of them the evolution of the system leads to the emergence of self-organized forms. The rich, unpredictable, beautiful dynamics of the global forms arising in the various experiments is related to the model used for the cell of the 3D system. The chaotic behavior of each cell, represented by the beauty of the strange attractor, is reflected into the beauty of the 3D forms like those shown in the gallery of reported models.

The evolutionary forms reported in the various graphs are isosurfaces that have been obtained time by time in the 3D space defined by the spatial coordinates i, j, k.

As concerns the cell dynamics $f(\mathbf{x}_{ijk})$, different chaotic laws have been simulated. In particular, the Lorenz system , the Rössler system [Strogatz (2003)] and the Chua's circuit [Chua (1998)] have been investigated. Moreover, the chaotic dynamics of several neuron models have been taken into account in order to emulate the global behavior of neural networks in a 3D space.

1.4.1 *Initial conditions*

Different initial conditions for the various experiments have been chosen and will be discussed in details in the following. However the idea underlying this choice is common to all the examples and will be briefly introduced in this section.

Initial conditions $\mathbf{x}_0(i, j, k)$ have been created starting by some topological form. In particular, they can be viewed as the composition of two functions σ and γ, describing the topological form taken into account.

Let us focus on a 3D-CNN made of third-order nonlinear units (as in most cases investigated in this paper) and let us indicate with i, j, k the three coordinates of the 3D space and $\hat{i} = \sqrt{-1}$.

Thus, the two functions σ and γ can be defined as follows:

$$\gamma : \mathbb{R}^3 \supseteq \Omega \rightarrow \mathbb{C}$$
$$\sigma : \mathbb{C} \rightarrow \mathbb{R}^3$$

and initial conditions are given by the following relation:

$$x_0(i, j, k) = \sigma \circ \gamma$$

The definition of the function γ makes use of complex curves used for the study of topological forms. For instance, a $(m : n)$ torus knot can be represented by the following equations

$$Z = \gamma(i, j, k) = \left(\frac{d^2 - 1}{1 + d^2} + \hat{i}\frac{2k}{1 + d^2}\right)^m - \left(\frac{2i}{1 + d^2} + \hat{i}\frac{2j}{1 + d^2}\right)^n$$

with $d = \sqrt{i^2 + j^2 + k^2}$. Then, initial conditions for the state variables x, y, z of each nonlinear unit of the 3D-CNN are created from equations (1.8) through the relations:

$$x_0(i, j, k) = A_x \mathrm{Re}(Z) + B_x$$

$$y_0(i, j, k) = A_y \mathrm{Im}(Z) + B_y$$

$$z_0(i, j, k) = A_z \mathrm{Re}(Z) + B_z$$

where A_x, A_y, A_z, B_x, B_y and B_z are real constants used to scale the initial conditions to match the dynamic range of the nonlinear units constituting the 3D-CNN.

1.4.2 *3D waves in homogeneous and unhomogeneous media*

First of all, let us consider a 3D-CNN, where each cell is a second-order nonlinear system, implementing a reaction-diffusion . Two examples are discussed. The first deals with an homogeneous medium, the second is an example of unhomogeneous medium.

The equations of the generic second-order cell C_{ijk} are the following:

$$\begin{cases} \dot{x}_{i,j,k;1} = k(-x_{i,j;1} + (1+\mu+\varepsilon)y_{i,j;1} - sy_{i,j;2} + i_1 + D_1(y_{i-1,j,k;1} \\ \quad +y_{i+1,j,k;1} + y_{i,j-1,k;1} + y_{i,j+1,k;1} + y_{i,j,k-1;1} + y_{i,j,k-1;1} - 6y_{i,j,k;2})) \\ \dot{x}_{i,j,k;2} = k(-x_{i,j;2} + (1+\mu-\varepsilon)y_{i,j;2} + sy_{i,j;1} + i_2 + D_2(y_{i-1,j,k;2} \\ \quad +y_{i+1,j,k;2} + y_{i,j-1,k;2} + y_{i,j+1,k;2} + y_{i,j,k-1;2} + y_{i,j,k-1;2} - 6y_{i,j,k;2})) \end{cases} \tag{1.8}$$

The parameters of the CNN cell have been chosen according to $\mu = 0.7$, $s = 1$, $i_1 = -0.3$, $i2 = 0.3$, $D1 = 0.1$, $D2 = 0.1$. For these values the 3D-CNN behaves as a nonlinear medium in which autowaves propagate. The parameter k accounts for the possible unhomogeneity of the medium.

In the first example, all the cells in the 40x40x40 3D-CNN have the same values of k ($k = 1$). In the center of the 3D-CNN, there is a "pacemaker" cell whose outputs are fixed to the values $y_{20,20,20;1} = 1$ and $y_{20,20,20;2} = -1$. This cell elicits the generation of autowaves in the neighboring cells. Fig. 1.5 shows the behavior of a 3D-CNN made of cells represented by equations (1.8). As it can be noticed, when two wavefronts collide they annihilate each other.

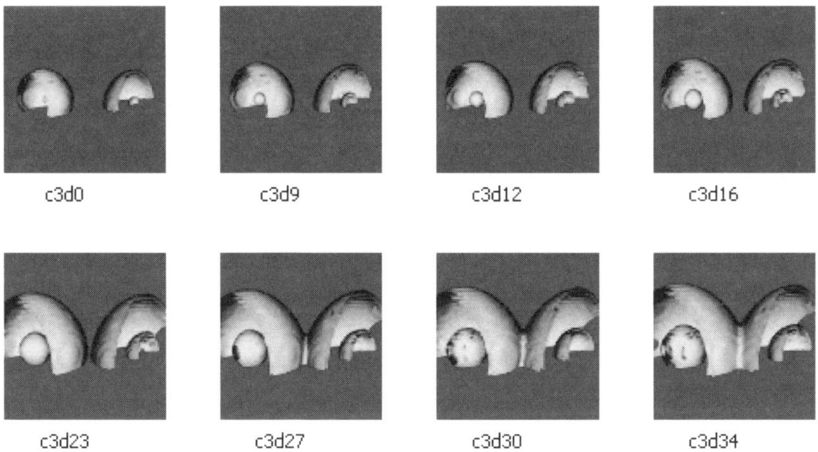

Fig. 1.5 Behavior of a 3D-CNN generating autowaves in a homogeneous medium.

In the second example, the cells of the 3D-CNN have different values of the parameter k. The value $k = 0.6$ characterizes "slow" cells, while the value $k = 1$ characterizes "fast" cells as schematically shown in Fig. 1.6(a). The whole CNN consists of 41x41x41 cells. Simulation results are shown in Fig. 1.6. Fig. 1.6(a) shows the initial configuration: we simulated an initial point of excitation (indicated by an arrow) and a "wall" in an unhomogeneous 3D medium. Figs. 1.6(b)-(d) represent the evolution of the RD-CNN. The presence of unhomogeneity in the medium clearly leads to the emergence of spiral waves. This experiment represents a fascinating emergent behavior shown by a complex system.

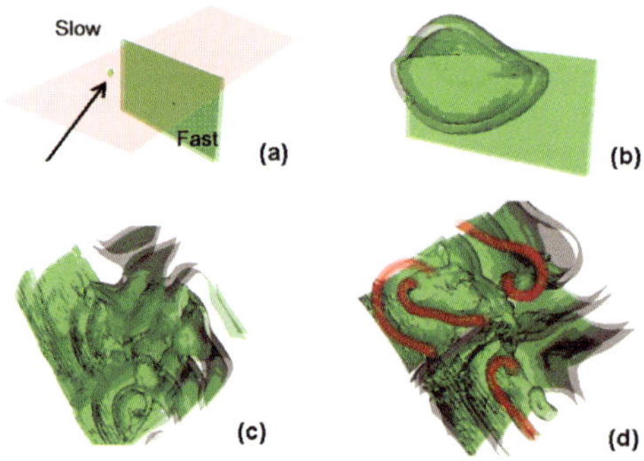

Fig. 1.6 Behavior of a 3D-CNN generating spiral wave in an unhomogeneous medium.

1.4.3 *Chua's circuit*

This example deals with the Chua's circuit [Chua (1998)]. In the case of a 3D-CNN made of Chua's circuits, equations (1.8) can be rewritten as follows:

$$\begin{aligned} \dot{x}_{ijk} &= \alpha(y_{ijk} - h(x_{ijk})) + D\nabla^2_{ijk}x \\ \dot{y}_{ijk} &= x_{ijk} - y_{ijk} + z_{ijk} \\ \dot{z}_{ijk} &= \beta y_{ijk} \end{aligned} \tag{1.9}$$

where

$$h(x) = 0.5((s_1+s_2)x+(s_0-s_1)(|x-B_1|-|B_1|)+(s_2-s_0)(|x-B_2|-|B_2|))+\varepsilon$$

and the diffusion term only acts on the first state variable $x_{ijk}(t)$. An array of 80x80x80 chaotic units has been considered, i.e. $1 \leq i, j, k \leq 80$

in equations (1.9). The parameters of each single unit have been chosen according to $\alpha = 9$, $\beta = 30$, $s_0 = -1.7$, $s_1 = s_2 = -1/7$, $\varepsilon = -1/14$, $B_1 = -1$ and $B_2 = 1$, in order to set the Chua's circuit in the bistability region. The diffusion coefficient has been fixed to the value $D = 0.1$. In order to visualize the behavior of the whole 3D-CNN, we considered an isosurface defined by $x_{ijk} = 0.1$. The emergent behavior leads to the formation of a form evolving in time in a non-repetitive way. Fig. 1.7 shows some frames of the evolution of a 3D-CNN made of Chua's circuits, where the formation of shapes and structures evolving in time is evident.

Fig. 1.7 refers to the following initial conditions:

$$z = \gamma(i,j,k) = \left(\frac{d^2-1}{1+d^2} + \hat{i}\frac{2k}{1+d^2}\right)^7 - \left(\frac{2i}{1+d^2} + \hat{i}\frac{2j}{1+d^2}\right)^5$$

$$\sigma(z) = (4\text{Re}z - 1.25, \text{Im}z, 4\text{Re}z + 1.25)$$

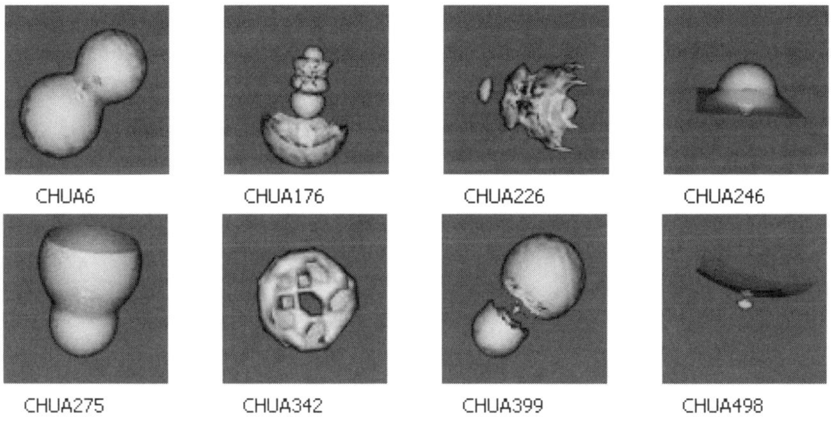

Fig. 1.7 Forms obtained by a 3D-CNN made of Chua's circuits.

Moreover, the stretching and folding dynamics appears in the frames shown in Fig. 1.8.

1.4.4 *Lorenz system*

The example reported in Fig. 1.9 deals with a CNN made of 60x60x60 chaotic Lorenz systems described by the following equations [Strogatz

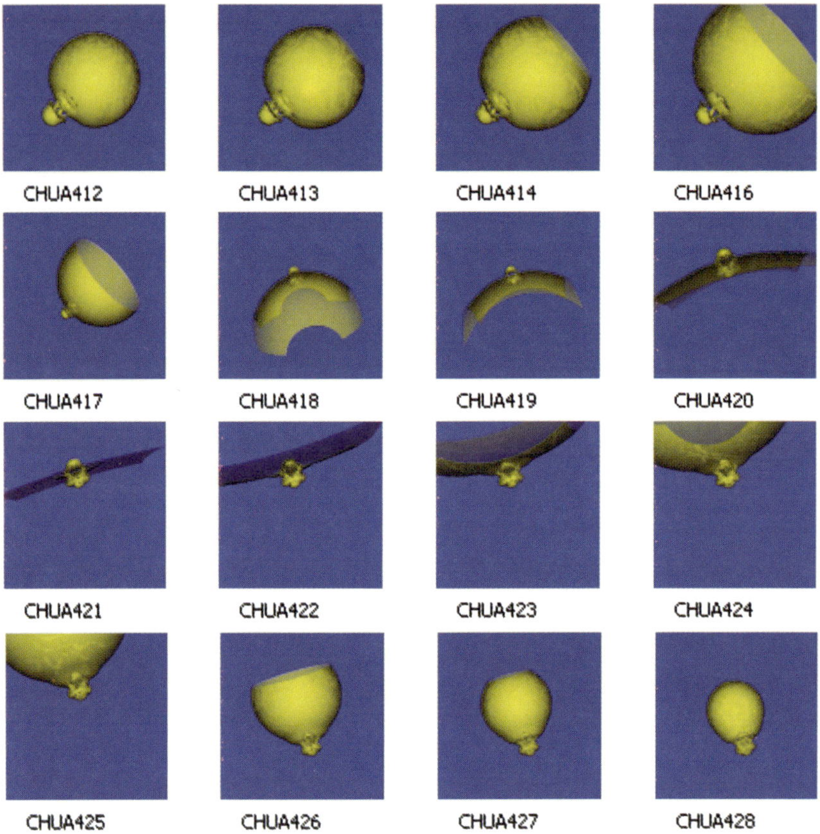

Fig. 1.8 Stretching and folding dynamics in 3D-CNN made of Chua's circuits.

(2003)]:

$$
\begin{aligned}
\dot{x}_{ijk} &= \sigma(y_{ijk} - x_{ijk}) + D\nabla^2_{ijk}x \\
\dot{y}_{ijk} &= rx_{ijk} - y_{ijk} + x_{ijk}z_{ijk} + D\nabla^2_{ijk}y \\
\dot{z}_{ijk} &= x_{ijk}y_{ijk} - bz_{ijk}
\end{aligned}
\tag{1.10}
$$

where the parameters have been chosen according to $\sigma = 10$, $r = 28$, $b = 8/3$ in order to set the well-known butterfly attractor. The diffusion coefficient has been fixed to the value $D = 0.5$.

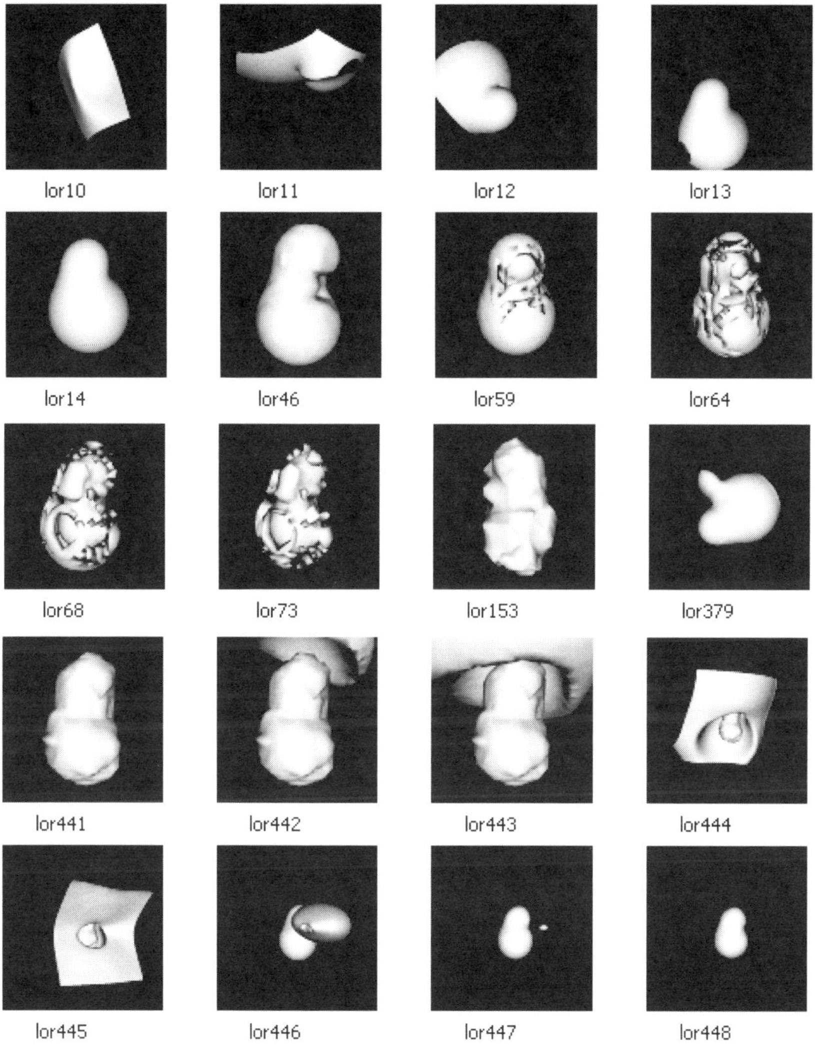

Fig. 1.9 Evolution of the isosurface $x_{i,j,k} = 2$ generated by a 3D-CNN made of Lorenz systems.

Initial conditions have been chosen as follows:

$$z = \gamma(i,j,k) = \left(\frac{d^2-1}{1+d^2} + \hat{i}\frac{2k}{1+d^2}\right)^6 - \left(\frac{2i}{1+d^2} + \hat{i}\frac{2j}{1+d^2}\right)^9$$

$$\sigma(z) = (4\mathrm{Re}z - 1.25, \mathrm{Im}z, 4\mathrm{Re}z + 1.25)$$

Fig. 1.9 shows the isosurface defined by $x_{ijk} = 2$.

Even in this case the evolution of the system leads to ever changing regular forms.

1.4.5 *Rössler system*

The emergence of organized forms and structures has been also observed in a 3D-CNN of Rössler units [Strogatz (2003)] as follows:

$$\begin{aligned}\dot{x}_{ijk} &= -y_{ijk} - z_{ijk} + D\nabla^2_{ijk}x \\ \dot{y}_{ijk} &= x_{ijk} + ay_{ijk} \\ \dot{z}_{ijk} &= b + x_{ijk}z_{ijk} - cz_{ijk}\end{aligned} \tag{1.11}$$

where the diffusion term only acts on the first state variable $x_{ijk}(t)$. An array of 60x60x60 chaotic units has been considered. The parameters of each single unit have been chosen according to $a = 0.2$, $b = 0.2$, $c = 5$, in order to set the Rössler chaotic attractor. The diffusion coefficient has been fixed to the value $D = 0.1$. In order to visualize the behavior of the whole 3D-CNN, we considered a level surface defined by $x_{ijk} = 0$. Fig. 1.10 shows some frames of the evolution of a 3D-CNN made of Rössler systems, where it is evident the formation of forms and structures evolving in time.

Initial conditions have been chosen as follows:

$$\begin{aligned} z &= \gamma(i,j,k) \\ &= \left(\frac{d^2-1}{1+d^2} + \hat{i}\frac{2k}{1+d^2}\right)\left(\frac{2i}{1+d^2} + \hat{i}\frac{2j}{1+d^2}\right) + \left(\frac{2i}{1+d^2} + \hat{i}\frac{2j}{1+d^2}\right)^5 \\ &\quad - \left(\frac{d^2-1}{1+d^2} + \hat{i}\frac{2k}{1+d^2}\right)^3 \end{aligned}$$

$$\sigma(z) = (4\text{Re}z - 0.25, \text{Im}z, 4\text{Re}z + 0.25)$$

1.4.6 *FitzHugh-Nagumo neuron model*

The first neuron model investigated is the FitzHugh-Nagumo (FHN) model [FitzHugh (1961); Nagumo (1960)] of spiking neurons described by the following equations:

$$\begin{aligned}\dot{v}_{ijk} &= \varepsilon v_{ijk}(1 - v_{ijk})\left(v_{ijk} - \frac{u_{ijk}+b}{a}\right) + D\nabla^2_{ijk}v \\ \dot{u}_{ijk} &= v_{ijk} - u_{ijk}\end{aligned} \tag{1.12}$$

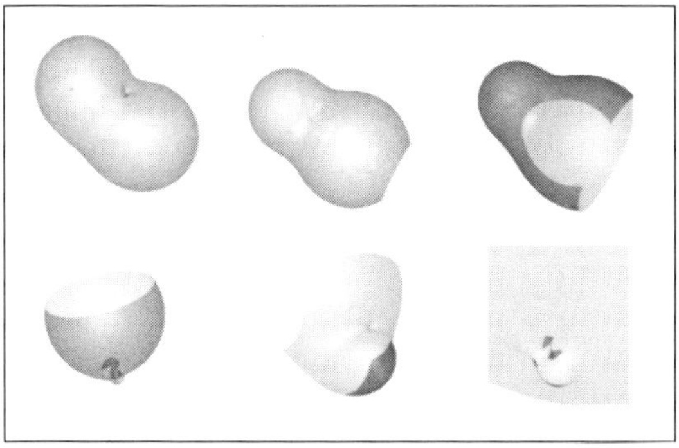

Fig. 1.10 Forms generated by a 3D-CNN made of Rössler systems.

The diffusion term only acts on the first variable.

The parameters have been chosen according to: $a = 0.75$, $b = 0.01$, $\varepsilon = 50$. An array of 50x50x50 neurons (1.12) coupled with a diffusion coefficient $D = 1$ has been taken into account. Some frames of the evolution of the isosurface defined by $x_{ijk} = 0.5$ are shown in Fig. 1.11.

1.4.7 *Hindmarsh-Rose neuron model*

A 3D-CNN where the basic cell is the Hindmarsh-Rose model [Rose (1989)] of bursting neurons is discussed here. The dynamics of this model is described the following equations:

$$\begin{aligned} \dot{x}_{ijk} &= y_{ijk} + ax_{ijk}^2 - x_{ijk}^3 - z_{ijk} + I + D\nabla_{ijk}^2 x \\ \dot{y}_{ijk} &= 1 - bx_{ijk}^2 - y_{ijk} \\ \dot{z}_{ijk} &= r(S(x_{ijk} - x_c) - z_{ijk}) \end{aligned} \tag{1.13}$$

where a diffusion term acting on the first variable has been included.

The parameters have been chosen according to: $a = 3$, $b = 5$, $r = 0.0021$, $S = 4$, $x_c = -1.6$, and $I = 0$ and 30x30x30 neurons (1.13) have been coupled with a diffusion coefficient $D = 0.75$ leading to very interesting results. Some frames of the evolution of the isosurface defined by $x_{ijk} = 0.5$ are shown in Figs. 1.12 and 1.13.

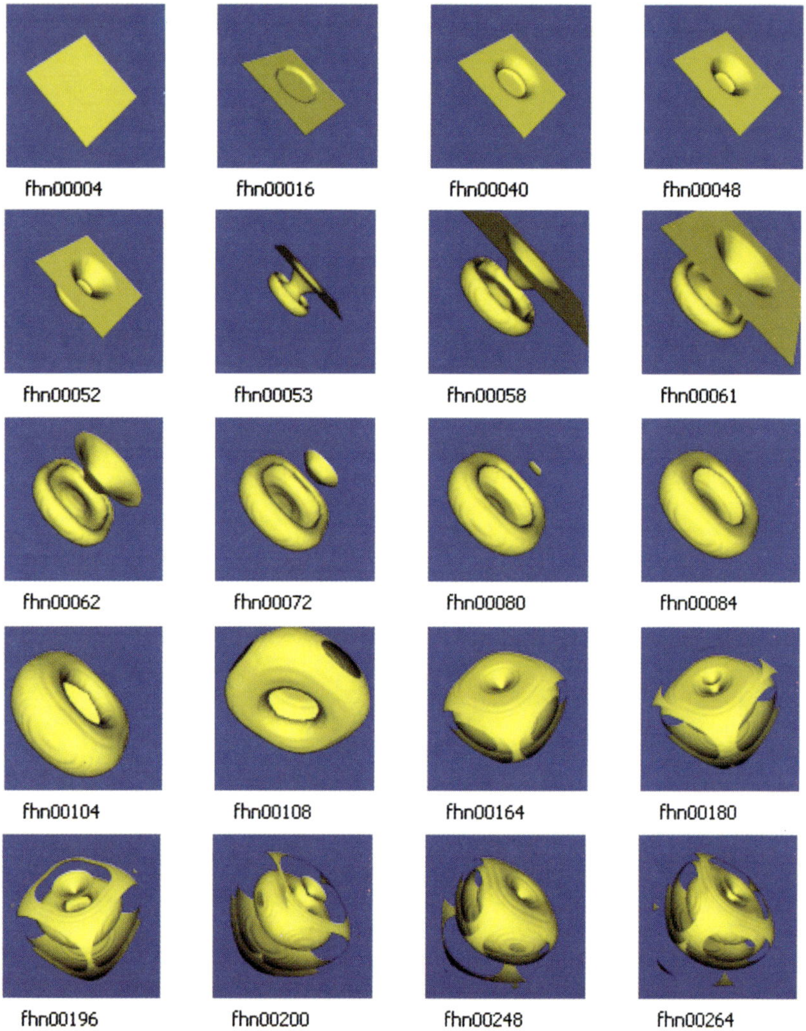

Fig. 1.11 Shapes generated by a 3D-CNN made of FHN neurons.

1.4.8 *Inferior-Olive neuron model*

This neuron model was proposed in [Giaquinta (2000)] to mimic the behavior of Inferior-Olive (IO) neurons. They are characterized by subthreshold oscillations. The dimensionless equations describing this model are the

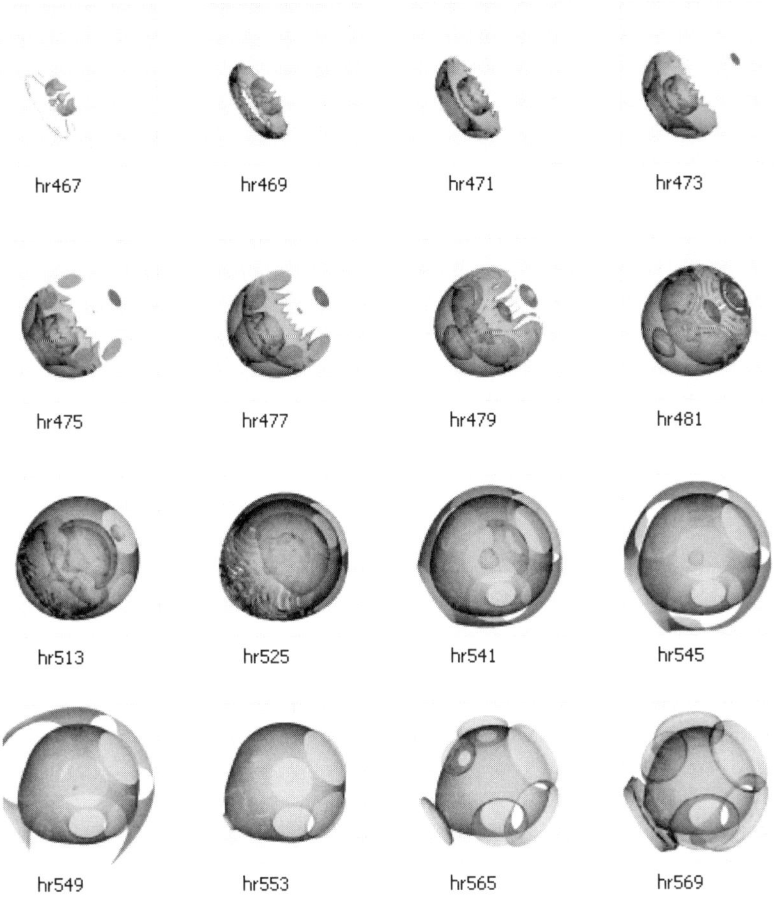

Fig. 1.12 Frames of the evolution of a 3D-CNN made of HR neurons.

following:

$$
\begin{aligned}
\dot{x}_{ijk} &= \frac{x_{ijk}(x_{ijk}-\gamma)(1-x_{ijk})-y_{ijk}}{\varepsilon}+D\nabla^2_{ijk}x \\
\dot{y}_{ijk} &= -\Omega z_{ijk}+r_{ijk}(A-z^2_{ijk}-r^2_{ijk})+D\nabla^2_{ijk}y \\
\dot{z}_{ijk} &= \Omega r_{ijk}+z_{ijk}(A-z^2_{ijk}-r^2_{ijk})+D\nabla^2_{ijk}z
\end{aligned}
\tag{1.14}
$$

with $r_{ijk}=\frac{y_{ijk}}{M}-x$.

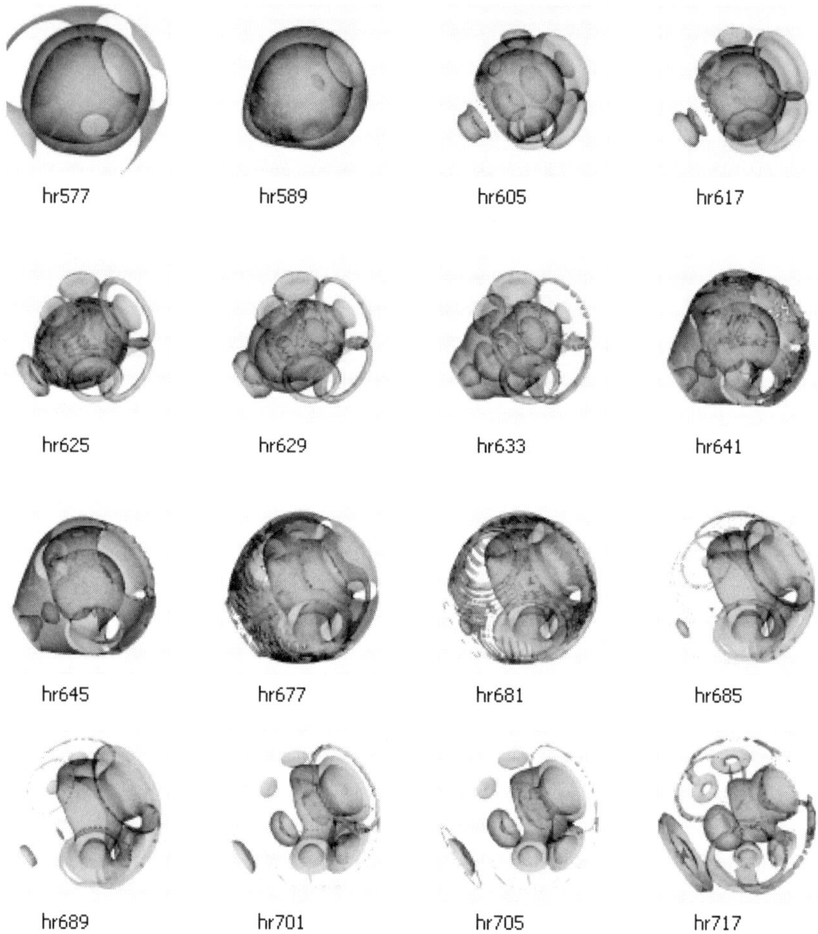

Fig. 1.13 Frames of the evolution of a 3D-CNN made of HR neurons.

The parameters have been chosen according to: $\varepsilon = 0.01$, $\gamma = 0.2$, $M = 0.5$, $A = 0.0006$, $\Omega = -1.6$. The CNN consists of 40x40x40 neurons (1.14) coupled with a diffusion coefficient $D = 0.001$. Some frames of the evolution of the isosurface defined by $x_{ijk} = -0.15$ are shown in Figs. 1.14.

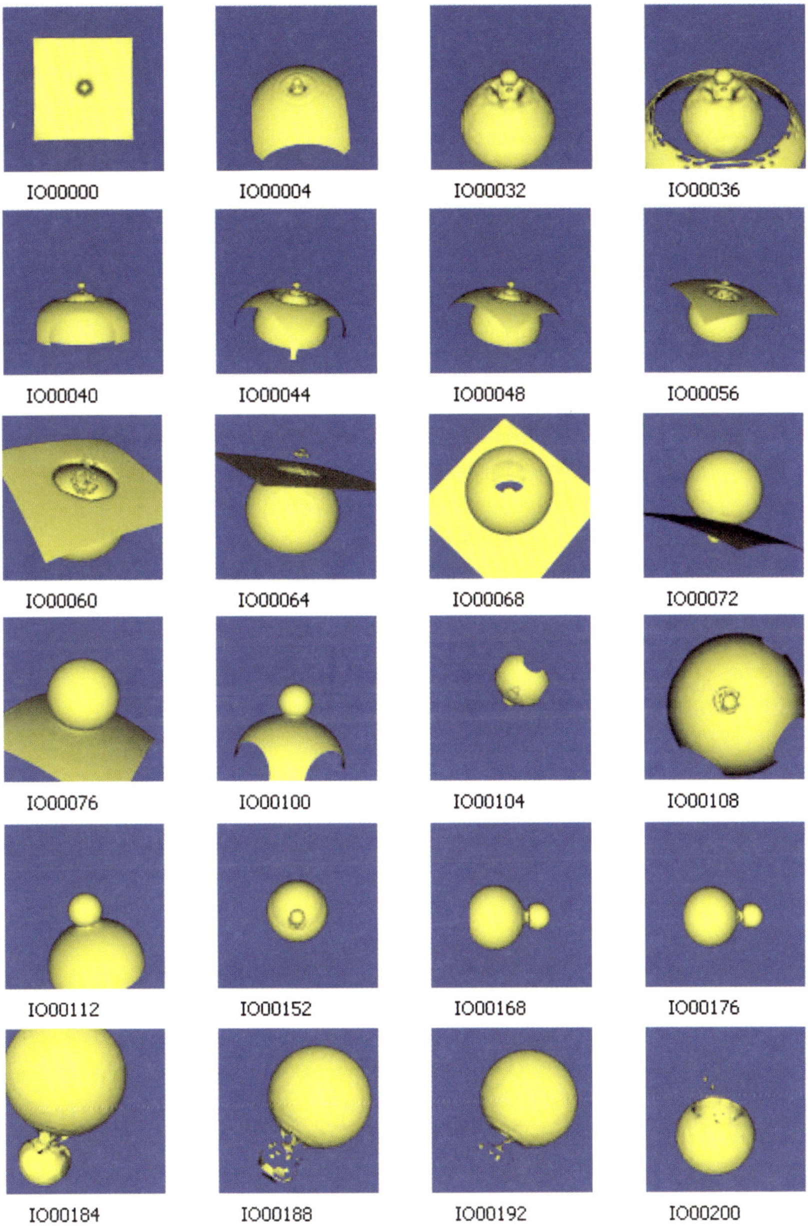

Fig. 1.14 Frames of the evolution of a 3D-CNN made of IO neurons.

Initial conditions have been chosen as follows:

$$z = \gamma(i,j,k) = \left(\frac{d^2-1}{1+d^2} + \hat{i}\frac{2k}{1+d^2}\right)^{7-2\frac{d^2}{D^2}} - \left(\frac{2i}{1+d^2} + \hat{i}\frac{2j}{1+d^2}\right)^{5+3\frac{d^2}{D^2}}$$

$$\sigma(z) = (2\text{Re}z - 0.4\frac{d^2}{D^2}, \text{Im}z + 0.4, \sin((\text{Re}z)^2 - (\text{Im}z)^2)$$

with $d = \sqrt{i^2+j^2+k^2}$ and $D = \sqrt{M^2+N^2+P^2}$.

1.4.9 *Izhikevich neuron model*

A recent neuron model was proposed by Izhikevich in order to conjugate accuracy of the model and computational resources needed to simulate large arrays of neurons [Izhikevich (2003)]. The model accounts both for different spiking behaviors (tonic, phasic and chaotic spiking) and for bursting behavior, depending on the parameters chosen. It can be described by the following equations [Izhikevich (2003)]:

$$\begin{aligned} \dot{v}_{ijk} &= 0.04v_{ijk}^2 + 5v_{ijk} + 140 - u_{ijk} + I + D\nabla_{ijk}^2 v \\ \dot{u}_{ijk} &= a(bv_{ijk} - u_{ijk}) \end{aligned} \tag{1.15}$$

with the spike-resetting

$$\texttt{if } v \geq 30\text{mV}\texttt{, then} \begin{cases} v \leftarrow c \\ u \leftarrow d \end{cases}$$

v and u are dimensionless variables, and $a = 0.2$, $b = 2$, $c = -56$, $d = -16$, and $I = -99$ are the parameters (chosen to set a chaotic spiking activity [Izhikevich (2004)]). The simulation of the 3D-CNN made of Izhikevich neurons has been carried out by considering 30x30x30 units, $D = 0.01$ and an isosurface defined by $v_{ijk} = -65.4$.

Initial conditions have been chosen as follows:

$$(v_{ijk}(0), u_{ijk}(0)) = \begin{cases} (-56, -112) \text{ if } rr < 5 \text{ AND } i > 0 \\ (20, 40) \text{ if } rr < 9 \text{ AND } i < 0 \\ (0, 0) \text{ otherwise} \end{cases}$$

with $r = \sqrt{i^2+j^2+k^2}$.

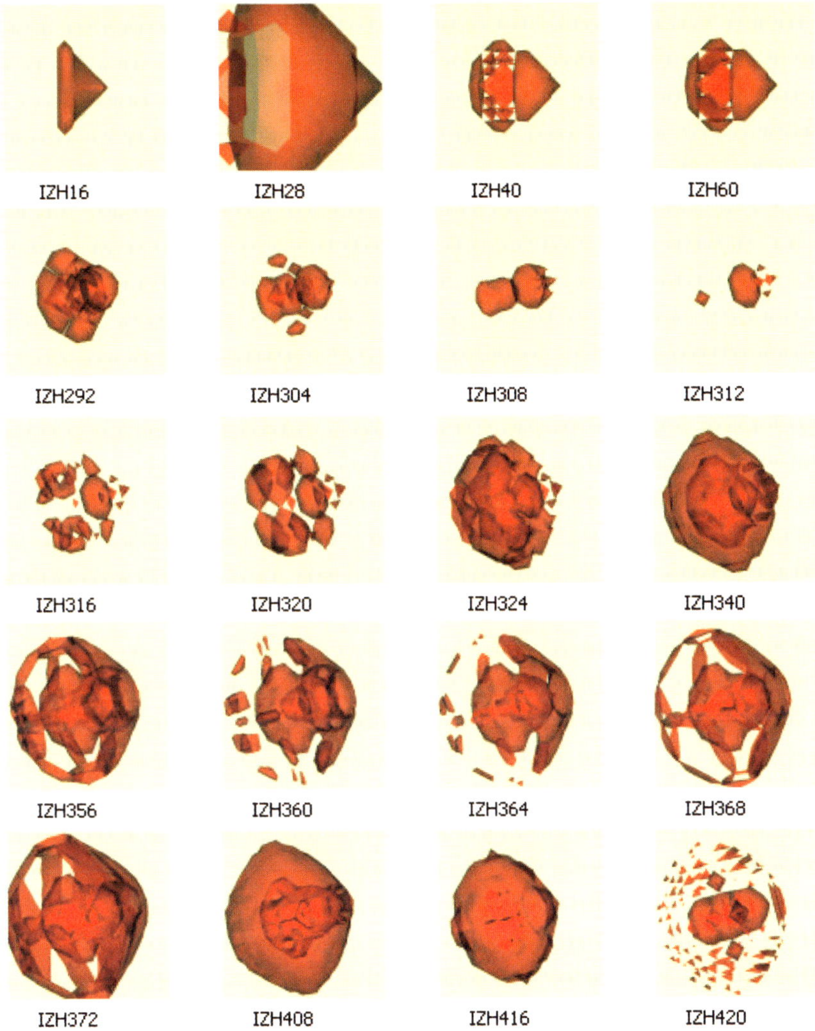

Fig. 1.15 Shapes generated by a 3D-CNN made of Izhikevich neurons.

1.4.10 *Neuron model exhibiting homoclinic chaos*

Another experiment has been carried on by using a cell model based on a CO_2 laser model. This model shows Shilnikov chaos and can be also taken as representative of a class of neuron dynamics with chaotic inter-spike intervals. The following dimensionless equations describe the behavior of this model:

$$
\begin{aligned}
\dot{x}_{1,ijk} &= k_0 x_{1,ijk}(x_{2,ijk} - 1 - k_1 \sin^2(x_{6,ijk})) + D\nabla^2_{ijk} x_1 \\
\dot{x}_{2,ijk} &= -\Gamma_1 x_{2,ijk} - 2k_0 x_{1,ijk} x_{2,ijk} + \gamma x_{3,ijk} + x_{4,ijk} + P_0 + D\nabla^2_{ijk} x_2 \\
\dot{x}_{3,ijk} &= -\Gamma_1 x_{3,ijk} + x_{5,ijk} + \gamma x_{2,ijk} + P_0 \\
\dot{x}_{4,ijk} &= -\Gamma_2 x_{4,ijk} + \gamma x_{5,ijk} + z x_{2,ijk} + z P_0 \\
\dot{x}_{5,ijk} &= -\Gamma_2 x_{5,ijk} + z x_{3,ijk} + \gamma x_{4,ijk} + z P_0 \\
\dot{x}_{6,ijk} &= -\beta x_{6,ijk} + \beta B_0 - \beta \frac{R x_{1,ijk}}{1 + \alpha x_{1,ijk}}
\end{aligned}
\tag{1.16}
$$

Parameters have been chosen according to [Pisarchik (2001); Ciofini (1999)] as follows: $R = 220$, $k_0 = 28.5714$, $k_1 = 4.5556$, $\Gamma_1 = 10.0643$, $\Gamma_2 = 1.0643$, $\gamma = 0.05$, $z = 10$, $\beta = 0.4286$, $\alpha = 32.8767$, $P_0 = 0.016$, $B_0 = 0.133$. For this set of parameters, homoclinic chaos appears.

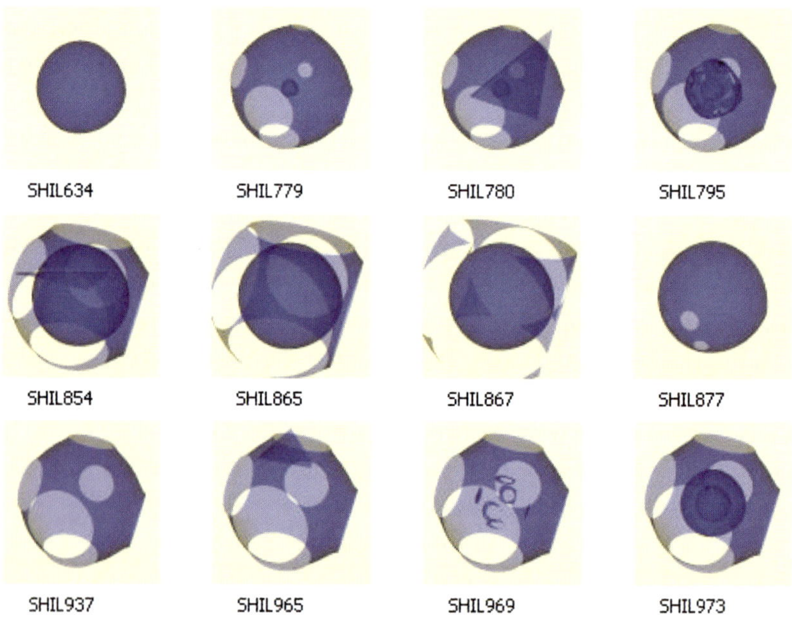

Fig. 1.16 Frames of the evolution of a 3D-CNN made of neurons (1.16) with homoclinic chaos.

We considered an array of 30x30x30 units diffusively connected with $D = 0.01$. Fig. 1.16 shows some frames of the evolution of a 3D-CNN made of units (1.16) starting from initial conditions chosen as follows:

$$\mathbf{x}_{ijk}(0) = \begin{cases} \mathbf{k_1} \text{ if } rr < 5 \text{ AND } i > 0 \\ \mathbf{k_2} \text{ if } rr < 9 \text{ AND } i < 0 \\ \mathbf{k_3} \text{ otherwise} \end{cases} \tag{1.17}$$

where $r = \sqrt{i^2 + j^2 + k^2}$ and $\mathbf{k_1}$, $\mathbf{k_2}$ and $\mathbf{k_3}$ are vectors of six constants. The isosurface shown in Fig. 1.16 is defined by $x_{1,ijk} = 403 * 10^{-3}$.

The movies of the experiments discussed above can be downloaded from the webpage www.scg.dees.unict.it/activities/complexity/3dcnn.html.

1.5 Conclusions

First of all a general remark regarding the previous experiments must be done: shapes in 3D-CNNs are the fingerprint of emergent phenomena . This occurs for all the adopted 3D-CNN configurations. Moreover, in the considered cases, the dynamical chaotic behavior of each cell leads to harmonic shapes in the 3D-CNN configuration. In order to reinforce the previous remark, let us consider a counterexample. Let us consider a system in which each cell is a random generator with a given probability distribution and let us consider the same coupling diffusive laws and grid dimension of the other experiments. Fig. 2.23 shows the results obtained by simulating such a system. In this case the level surface is irregular and there is no clear form arising. Regularity, self-organization is not possible in this case.

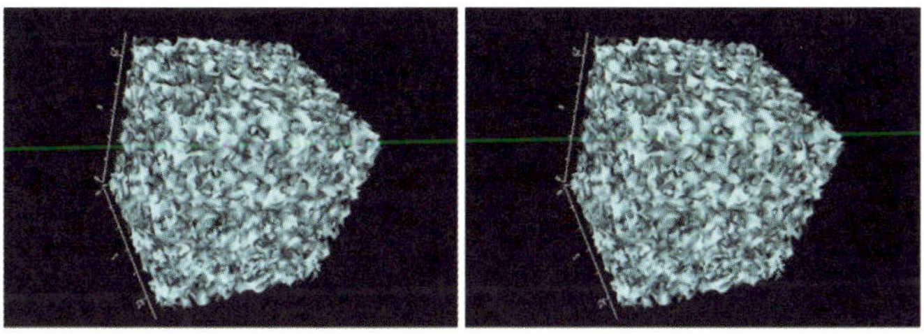

Fig. 1.17 When random generators are coupled together into a 3D-CNN, there is no self-organization and regular shapes do not form.

Let us consider the various shape trends shown in the various sequences of the previous Section, the following strong observation can be done: each shape form is not recurrent in time. Moreover, in many of them, spatial symmetries in each frame are evident. The variety of shapes are related to the cell dynamics: however some of them reflect global time-related features like the stretching and folding phenomenon.

Following the introduction of the emerging shape generation phenomena, the complex evolution trend by using 3D-CNN dynamics and the role of self-organization in the previous phenomena, the problem to try to discover recurrent patterns both in the 3D-CNNs shapes and in modern art is faced. It is not the aim of this paper to investigate on computer based arts or to dealt with the well known evolutionary art [Bentley (2001)]. This is to remark the value of the staggering complexity we are dealing with. Only few examples are here reported. Let us consider the Mirò paint "Still Life with Old Shoe"; recurrent patterns of it are found in the 3D-CNN generated shape when a Lorenz system is adopted as cell unit. This is shown in Fig. 1.18.

Fig. 1.18 A form generated by the Lorenz 3D-CNN and the paint "Still Life with Old Shoe" by Mirò (black and white reproduction of the original paint [Miró (1937)]).

Let us consider now the Robert Delaunay's painting study. He started as an impressionist under the suggestion of Cezanne and taking into account the Cubism, started an analytical research on the form in relationship to the multiplication of light planes. An example of this study is given in the paint "Joie de vivre" where he expressed the emergence of light and

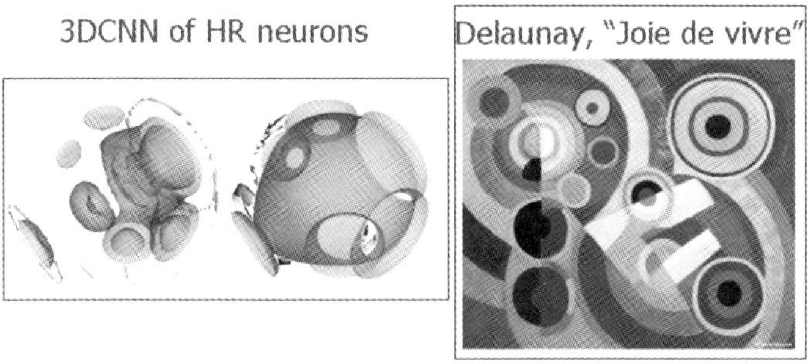

Fig. 1.19 Shapes generated by a 3D-CNN of HR neurons and "Joie de vivre", oil on canvas, by Delaunay [Delaunay (1930)].

nature by using a painter language made by the contrast of colors whose expressions are sequences of closed curves. Let us compare this paint with the 3D-CNN-generated surfaces as shown in Fig. 1.19. In this case the 3D-CNN cell is the HR dynamical system.

Each observed dynamics of frames included highly organized shapes. Even if each cell is chaotic, even if each cell is characterized by the geometrical form of the corresponding attractor with a defined shape, the 3D-CNNs complex patterns generate both an unusual complexity and an astonishing unpredictability. The discovery of recurrent patterns between 3D-CNN dynamics and artist paints remarks the emergent characteristics of both underlying organized complexities. Through examples like those, it is emphasized that the dynamical evolution of coupled cells can produce very beautiful and rich patterns drawn by artists.

Let us observe now the sequence of frames respectively reported in Fig. 1.12 and in Fig. 1.13. Regular line segments are evident, unpredictable ordered fragmentation explodes, a certain type of intermittency frequently appears. Moreover irregular shining flashes, like phosphenes appear in some of the frames. Moreover, in both the forms of Fig. 1.15 and Fig. 1.16 circular waves appear. These forms are particularly unstable showing very fast changes in the shape, in the size of them and in the time-scale evolution. The various circles run giving us an impressive image of vortices and turbulence.

The dynamical combination of the shapes give us a global view perception whose effect in our mind is much more the sum of the single shape contribution. We are dealing with a complex visual pattern generator. The information contained in the whole is many times greater then the sum of the information contained in the single parts. There exists a parallelism between the previous considered frames and those referred in the migraine aura or in general in the complex hallucination phenomena. A detailed description of these phenomena is widely reported in literature [Sacks (1992); Klüver (1967)]. Moreover the visual effect of hallucinations is considered as a complex emergent dynamical phenomenon [Dalhem (2003)]. In particular, many scientists view the hallucination as the propagation of Reaction Diffusion wave in neural tissue. Many models have been proposed in this direction. Cortical organization [Dalhem (2000)] in migraine aura is supposed and mathematical models introduced to explain analytically the phenomena are widely accepted. Ermentrout and Cowan [Ermentrout (1979)] modelled this phenomena by using the relationship between inhibitory and excitatory neurons. However they introduced a linearized model that, even if allowed to establish the parameter relationship for deriving the instability condition, did not express the emergent mechanism of the hallucinations due to the nonlinearity effects.

The myriad pattern formation due to the migraine aura is a fascinating phenomenon. The diversity of migraine auras in the various forms underlines its complexity. Moreover the phenomena emulated in the reported experiments have been generated by using 3D-CNNs in reaction-diffusion configuration using as cells integrated and fire neuron models. It is not the aim of this experiment to model by using a grid of cortical neurons; moreover the introduced 3D-CNN strategy allows to emulate real self-organizing phenomena in the visual cortex. The CNN model for the hallucinations has been also approached in [Chua (1998)]. The experiments reported in our paper regard a wider set of hallucination phenomena. They have been obtained thanks to the 3D-CNN architecture and to the introduction of more complex cell dynamics with respect to those used in previous papers.

In view of the appealing field of computational neuroscience and in particular in the area of computer simulation algorithms devoted to explore the nervous system, the reported examples reinforce the suitability to adopt 3D-CNN circuits for emulating brain emergent phenomena [Arena (2005)].

CNN exactly translates the meaning of complexity in term of electronic circuits. The results of coupling many simple nonlinear circuits give us a global circuit whose capabilities are much more than the predictable performances obtained by summing the single cell circuit contribution. This allows to create artificially emergent patterns. Moreover, the same emergent behavior has been discovered in the visual cortex phenomena of migraine

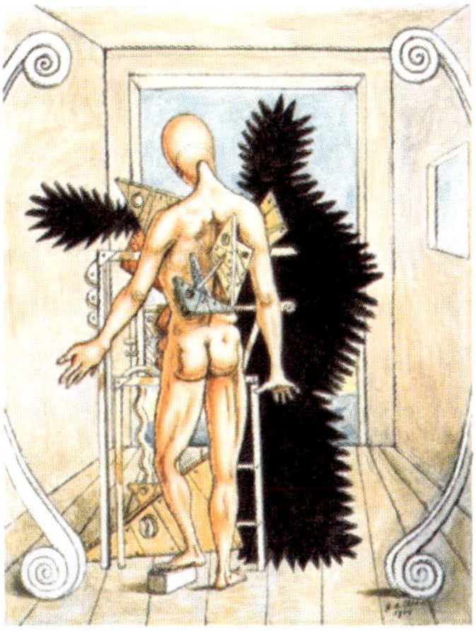

Fig. 1.20 "Il rimorso di Oreste" by Giorgio De Chirico.

aura and in the hallucination. The richness of forms and their combination reflect a further impressive example of complexity. A further remark: the metaphysical art represents another example of complexity from an aesthetic point of view. Thinking Complexity is a new point of view, a new methodology. The vision of complexity theme is resumed, as an example, in the famous paint of Giorgio De Chirico "Il rimorso di Oreste" shown in Fig. 1.20.

In this Chapter, the use of 3D-CNN generalized paradigm to generate sequences of emerging shapes and forms has been faced. A wide range of organized results from the evolution of 3D-CNN dynamical system has been shown. Different cells dynamics have been taken into account. Locally active cells or cells at the edge of chaos have been chosen in order to assure pattern formation. Complex 3D patterns emerged from the various experiments have been critically discussed. Links among circuits, art and neuroscience emerged thanks to the universality of CNN formalization.

In his impressive book "The Structure of Scientific Revolutions", Thomas Kuhn founded his theory on the concept of paradigm [Kuhn (1962)]. With this term Kuhn indicates the "scientific conquests universally accepted which, for a period, give a model for problems and solutions for people that made research in a particular field". When the paradigm changes, a critical breakpoint occurs in science and therefore occur the bifurcation conditions under which a new theory replaces an old one. In any case each scientific relationship is a set of ideas that lead us to a small or big settling when they are replaced into an old scientific paradigm.

Kuhn observed that the scientific revolutions could be small or big, but both of them have the same structure, the same characteristics when they occur. In fact what happens is like the same emergent phenomenon that occurs in the sand pile or in earthquake events. The cause could be small or big, but what is remarkable is the emergent behavior that arises after the paradigm is changed. In our opinion what Kuhn remarks for the scientific revolution occurred with the invention of CNN in the field of information technology.

The emergent behavior of a scientific revolution is related to the criticism of some aspects of classical or previously accepted paradigms: the positive criticism of scientists leads to introduce new paradigms in order to overcome problems not solved by previous theories. This is the starting point: ideas slowly evolve until the emergence occurs and the new paradigm switches to the revolution! In the history of distributed intelligence paradigms a fundamental limit of perceptron architectures has been highlighted by Minsky and Papert themselves in [Minsky (1969)] that established that a perceptron can not tell whether two labyrinthine patterns on the cover are connected or not. The CNN locally connected networks can solve such a problem! It has been proved that a locally connected network like CNN has the properties to recognize local functions [Chua (1998)]. Therefore in order to overcome a problem a new successful paradigm has been introduced. The critical point of a technological revolution has been established and now after sixteen years the change of the paradigm in the area of connectionism leads us to a scientific revolution! The revolution has been related to the new CNN proposed approach. Moreover the increase of interest in the CNN paradigm worked like an attractor. Starting from pattern recognition, complex model behaviors, vision, neuromorphic models, robotics, neuro-science problems have been faced in term of CNN paradigm. The new formalization allowed to conceive new advanced equipments and a revolution in the field of information technology is going. The dynamical richness of CNN based architecture allowed to reformulate classical problems in a new formalization. The last effort proposed in this paper is to investigate on the universality of the 3D-CNN in order to discover emergent 3D shapes. Self-organization that is the core of the emergent characterized CNN system allowed us to make shorter the distance between technological science, art and neuroscience.

References

P. Arena, M. Bucolo, S. Fazzino, L. Fortuna, M. Frasca, "The CNN Paradigm: Shapes and Complexity", *International Journal of Bifurcation and Chaos*, Vol. 15, No. 7, pp. 2063-2090, 2005.

Arena P., Baglio S., Fortuna L. and Manganaro G.(1996), "How State Controlled CNNs generate the dynamics of the Colpitts-Like Oscillator", *IEEE Trans. Circuits and Systems I*, vol. 43, no. 7, pp. 602-605.

Arena P., Baglio S., Fortuna L. and Manganaro G.(1995), "Chua's circuit can be generated by CNN cells", *IEEE Trans. Circuits and Systems I*, vol. 42, no. 2, pp. 123-125.

Arnheim R., *Art and Visual Perception: A Psychology of the Creative Eye*, University of California Press; 2nd Rev&En edition (November 1, 1974).

Bentley .J. and Corne D.W. (2001) *Creative Evolutionary Systems* Morgan Kaufmann Pub. July 2001.

Bogousslavsky J., (2003) "The Neurology of Art – The Example of Giorgio de Chirico", Eur. Neurol. 50:189-190.

Chua L. O., Roska T. (1993) "The CNN paradigm", *IEEE Trans. Circuits and Systems I*, 40, pp. 147–156.

Chua L. O., Yang L. (1988), "Cellular neural networks: theory", *IEEE Trans. Circuits and Systems I*, 35, pp. 1257–1272.

Chua L. O., Yang L. (1988), "Cellular neural networks: applications", *IEEE Trans. Circuits and Systems I*, 35, pp. 1273–1290.

Chua L.O., *CNN: A Paradigm for Complexity*, World Scientific Series on Nonlinear Science, Series A , Vol 31, 1998.

Ciofini M., Labate A., Meucci R. (1999), "Stabilization of unstable fixed points in the dynamics of a laser with feedback", *Physical Review E*, vol 60, no. 1.

Dahlem M. A., Engelmann R., Löwel S. and Muller S. C., (2000) "Does the migraine aura reflect cortical organization?, European Journal of Neuroscience, vol. 12, pp. 767-770.

Dalhem M.A. and Müller (2003) "Migraine aura dynamics after reverse retinotopic mapping of weak excitation waves in the primary visual cortex", *Biol. Cybern.*, 88, 419-424.

Dalí S. (1951) "Tête Raphaelesque Eclatée", Scottish National Gallery of Modern Art, Edinburgh.

Delaunay R. (1930) "Joie de vivre". Pompidou Center, Paris, France.

Duchamp M. (1947), "Prière de toucher". Modern Art Museum of the City of Paris, Paris, France.

Ermentrout G.B. and Cowan J.D. (1979) "A mathematical theory of visual hallucination patterns", *Biol. Cybern.*, 34, 137-150.

FitzHugh R. (1961), "Impulses and physiological states in theoretical models of nerve membranes", *1182-Biophys. J.*, 1, pp. 445–466.

Fortuna L.,Rizzo A., Xibilia M. G., (2003) "Modeling Complex Dynamics via Extended PWL-based CNNs", *Int. J. Bifur. Chaos*, Vol. 13, No. 11, pp. 3273-3286.

Giaquinta A., Argentina M., Velarde M. G. (2000), "A Simple Generalized Excitability Model Mimicking Salient Features of Neuron Dynamics", *Journal of Statistical Physics*, , Vol. 101, Nos. 1-2, 665-678.

Giles J., (2004) "Neuroscience: Change of mind", Nature 430, pp. 14 (01 Jul 2004).

Haken H., *Synergetics, An Introduction*, Springer Series in Synergetics, SpringerVerlag, Heidelberg, Berlin, New York, 1983.

Haken H., *Cooperative phenomena in systems far from equilibrium and in non-physical systems*, Rev. Mod. Phys. 47 67-121, spec. 89-91, 1975.

Izhikevich E. M., (2004), "Which Model to Use for Cortical Spiking Neurons?", *IEEE Transactions on Neural Networks*, Vol. 15, No. 5, 1063–1070.

Izhikevich E. M., (2003), "Simple Model of Spiking Neurons", *IEEE Transactions on Neural Networks*, Vol. 14, No. 6, 1569–1572.

Klüver H. (1967) *Mescal and the Mecanism of Hallucinations*, University Press of Chicago, Chicago.

Kuhn T. (1962) *The Structure of Scientific Revolutions*, University of Chicago Press.

Langton, *Computation at the Edge of Chaos: Phase Transition and Emergent Computation.*, Physica D, 42, pp.12-37, 1990.

Lloyd S., "Measures of complexity: a nonexhaustive list", *IEEE Control Syst. Mag.*, vol. 21, pp. 7-8, Aug. 2001.

Lynton R., *The Story of Modern Art*, Phaidon Press (January 1, 1994).

Minsky M. and Papert S. A. *Perceptrons: An Introduction to Computational Geometry.* MIT Press, Cambridge, MA, expanded edition, 1988/1969.

Mirò J. "Still Life with Old Shoe". 1937. The Museum of Modern Arts, New York, NY, USA.

Morales C. (1991) "Profile: Eladio Dieste" In MIMAR 41: Architecture in Development. London: Concept Media Ltd., 1991.

Nagumo J., Arimoto S., and Yoshizawa S. (1960), "An active pulse transmission line simulating 1214-nerve axons", *Proc. IRL*, 50, pp. 2061–2070.

Nicolis G. and Prigogine I., *Self-Organization in Non-Equilibrium Systems*, Wiley, 1977.

Orsucci F., *CHANGING MIND, Transitions in Natural and Artificial Environments*, World Scientific, 2002.

Otto F. (1982) "Frei Otto's New Work" 1982. In MIMAR 4: Architecture in Development. Singapore: Concept Media Ltd.

Pisarchik A. N., Meucci R., Arecchi F. T. (2001), "Theoretical and experimental study of discrete behavior of Shilnikov chaos in a CO2 laser", *The European physical journal D*, 13, 385-391.

Podoll K., Robinson D., Nicola U., (2001) "The migraine of Giorgio De Chirico. 1. History of illness." Neurol. Psychiatr. Brain. Res., 9:139-156.

Rose R.M. and Hindmarsh J.L., "The assembly of ionic currents in a thalamic neuron. I The three-dimensional model", *Proc. R. Soc. Lond. B*, 237, 267-288.

Sacks O. W., *Migraine*, Vintage Books, 1992.

Schrodinger, *What is Life? The Physical Aspect of the Living Cell*, Cambridge University Press, 1948.

Strogatz S. H., (2000), *Nonlinear Dynamics and Chaos*, Perseus Publishing.
Yang T., Chua L. O., "Testing for local activity and edge of chaos", *International Journal of Bifurcation and Chaos*, Vol. 11, No. 6, 1495-1591, 2001.
Wilkinson M., Robinson D. (1985) "Migraine art", Cephalalgia 5:151-157.
Zeki S., *Inner vision: An Exploration of Art and the Brain*, 1999.

Chapter 2

Emergent Phenomena in Neuroscience

L. Fortuna, M. Frasca and M. La Rosa

In this chapter, after a brief introduction on neuron models, the properties of several complex systems made of elementary units modelling neuron dynamics are examined. Electronic implementations of both single cell dynamics and groups of neurons are also considered.

2.1 Introductory Material: Neurons and Models

In literature there exist several mathematical models of neuronal dynamics observed in biological neurons. These neuron models are based on a system of differential nonlinear equations and reproduce the qualitative features of the dynamics of excitable/oscillatory processes. Usually, these models account for single cell behavior. Some of the main neuronal features that they reproduce are: excitability, threshold behavior, beating, bursting oscillations, bistability, hysteresis.

Among the most important characteristics of neuronal dynamics are the presence of a threshold and connected to this threshold the excitability of the neuron state. When excited from an external input (which models external current input), the neuron (and in particular the membrane potential) has two types of response. If the input is under the threshold, after excitation the membrane potential relaxes back to its resting equilibrium value. In this case the behavior is the typical behavior of a first-order system. If the input is over the threshold, a stereotyped response, called spike or potential action, arises. From the viewpoint of the theory of dynamical systems, this response is a large excursion in the phase plane.

Obviously, neurons do not respond at the same manner and even when

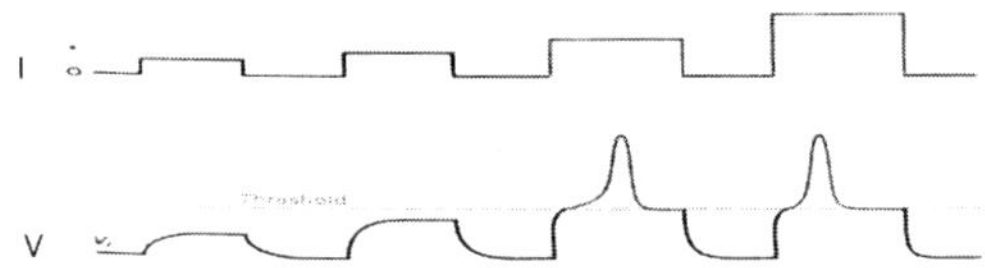

Fig. 2.1 Response v of an excitable system to different amplitude stimuli.

no input is applied, they can have different endogenous electrical patterns. There exist neurons which do not fire (in this case they are called silent), neurons which fire at constant frequency (the behavior is called beating), and neurons which generate regular burst of action potentials separated by hyperpolarization of the membrane (bursting behavior). These are the basic behaviors of neurons, but many other electrical patterns are observed in real data, for instance irregular bursting or chaotic spiking are quite common behaviors.

The complex features of neurons are regulated by different voltage-dependent channels which allows ions of different elements (for instance, Na, Ca and K) to enter or leave the region inside the membrane itself. We will not enter in the details of ionic channels, for which we refer to the specialized literature [Shepherd (1997)].

2.1.1 *Models of excitability*

Excitability can be modelled by taking into account simple considerations. From the viewpoint of neuron dynamics, the response of a neuron stimulated by injecting it with a pulse of current depends on the amplitude of the stimulus. If the stimulus is small, the membrane potential increases, then relaxes back to its resting potential. If the stimulus exceeds the threshold, the neuron fires (i.e. it produces a large voltage spike before returning to rest). In terms of the theory of dynamical systems this behavior matches the definition of an excitable system. A system is excitable if it has a unique, globally attracting rest state and a large enough stimulus can send the system on a long excursion through phase space before it returns to the resting state.

A very simple model of excitability is provided by the following dynamical system:

$$\dot{\theta} = \mu - \sin\theta \tag{2.1}$$

where θ is a variable with values in $[0, 2\pi]$. If $\mu < 1$ system (2.1) has two equilibrium points: one is stable, the other is unstable. Fig. 2.2 shows the phase diagram of the system. Initial conditions close to the unstable equilibrium point can follow two different trajectories. They can relax to the stable equilibrium point or they can follow a large excursion before reaching the stable equilibrium point, thus accounting for the mechanism of excitability.

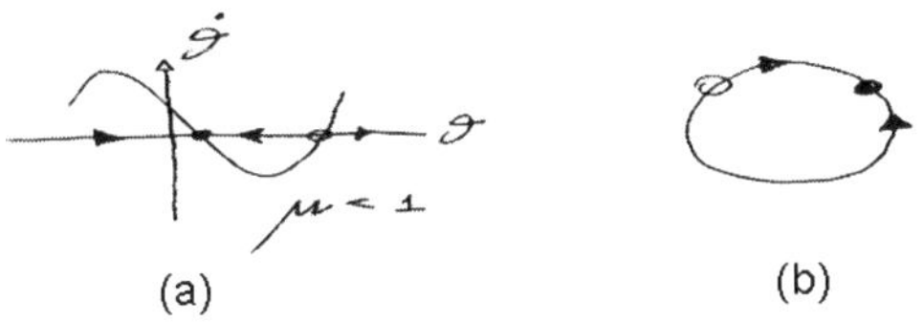

Fig. 2.2 Schematic illustration of neuron excitability.

2.1.2 *The Hodgkin-Huxley model*

The most complete model of neuron dynamics is the model of Hodgkin-Huxley. The core of this model is the equation for the membrane potential:

$$C\frac{dV}{dt} = -F + I_s + I \tag{2.2}$$

where V represents the membrane potential. F is the membrane current, I_s is the sum of external synaptic currents, and I is the external injected current. The membrane current is usually modelled as follows:

$$F(V, m, n, h) = g_L(V - V_L) + g_K n^4(V - V_K) + g_{Na} h m^3(V - V_{Na}) \tag{2.3}$$

where V_L, V_K and V_{Na} are reversal potentials and m, n and h conductance variables. Conductance variables are regulated by the following equations:

$$\tau_X(V)\frac{dX}{dt} = X_\infty(V) - X \tag{2.4}$$

where $X \in \{m, n, h\}$ represents one of the conductance variables. The functions $\tau_X(V)$ and $X_\infty(V)$ are obtained by fitting experimental data, thus they depend on the particular neuron which has to be modelled. Usually, the following form is assumed for these functions:

$$\begin{aligned} \tau_X(V) &= \frac{1}{\alpha_X(V)+\beta_X(V)} \\ X_\infty(V) &= \alpha_X(V)\tau_X(V) \end{aligned} \tag{2.5}$$

α_X and β_X depend on the experimental data. For instance, they may have the following expression:

$$\begin{aligned} \alpha_m(V) &= \frac{0.1(V+40)}{1-e^{-0.1(V+40)}} \\ \alpha_n(V) &= \frac{0.01(V+55)}{1-e^{-0.1(V+55)}} \\ \alpha_h(V) &= 0.07e^{-0.05(V+65)} \\ \beta_m(V) &= 4e^{-0.0556(V+65)} \\ \beta_h(V) &= \frac{1}{1+e^{-0.1(V+35)}} \end{aligned} \tag{2.6}$$

The Hodgkin-Huxley model is very complex and can fit a lot of experimental data. On the other hand, reduced mathematical models can capture the essential features of the system and do not require the computation effort needed to simulate the complete Hodgkin-Huxley model. One of the simplified model of neuron dynamics is the model introduced by FitzHugh and Nagumo.

2.1.3 *The FitzHugh-Nagumo model*

The FitzHugh-Nagumo model is described by the following equations:

$$\begin{aligned} \frac{dv}{dt} &= f(v) - w + I_a \\ \frac{dw}{dt} &= bv - \gamma w \end{aligned} \tag{2.7}$$

where $f(v) = v(a-v)(v-1)$. This model takes into account the excitability of the neuron (through a Hopf bifurcation). It is an example of a *relaxation oscillator*, i.e. an oscillator with slow-fast dynamics, where v is the fast variable and w the slow variable. In fact, v may adjust rapidly and maintain a pseudo-equilibrium along the stable branches of $f(v) - w + I_a = 0$.

To illustrate the mechanism for excitability, let us assume the following parameter values: $b = \gamma = 2 \cdot 10^{-3}$, $a = 0.5$. Let us first consider $I_a = 0$. The behavior of the FHN neuron under these conditions is shown in Fig. 2.3. After the emission of a spike, the neuron membrane potential relaxes back to its resting value. In this case the FHN neuron models an

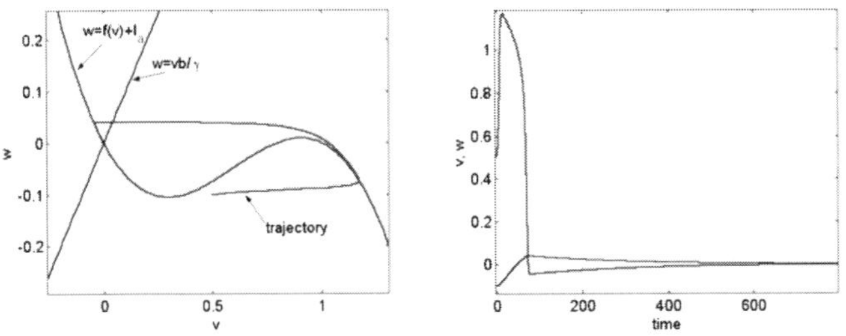

Fig. 2.3 FHN neuron with $I_a = 0$.

excitable system. When $I_a = 0.5$, another behavior emerges as shown in Fig. 2.4. In this case the behavior of the FHN neuron is beating. The FHN neuron periodically emits spikes (periodic behavior). The bifurcation of the behavior from a stable equilibrium point to a stable limit cycle is a subcritical Hopf bifurcation. It is interesting to note that most of the mechanisms for excitability in neuron models can be explained through only two bifurcations: the Hopf bifurcation illustrated by this example and the saddle-node limit cycle bifurcation. These two different mechanisms permit a classification of neuron models.

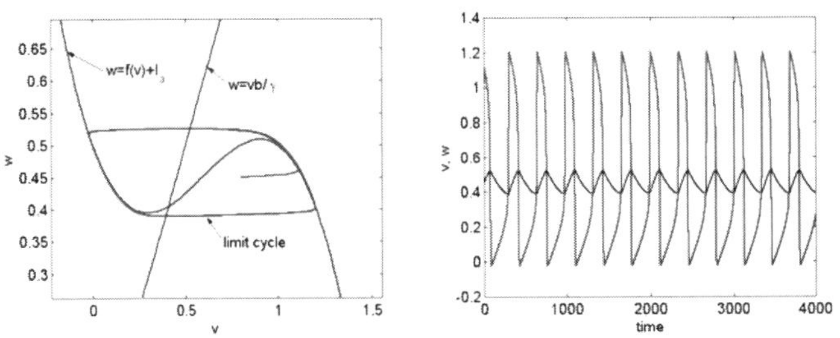

Fig. 2.4 FHN neuron with $I_a = 0.5$.

2.1.4 *Class I and class II excitability*

Neuron excitability and its characteristics depend on the bifurcation of the quiescent state. Two universal mechanisms are taken into account as bifurcation mechanisms for the quiescent state of many neuron models (see [Rinzel (1989)], and for a recent review [Izhikevich (2000)]), as schematically shown in Fig. 2.5. They lead to a classification of neural excitability which is in accordance with the classification proposed by Hodgkin [Hodgkin (1984)] on the basis of the different types of response observed in neurons. According to this classification, class I neurons generate action potential with arbitrarily low frequency and their frequency increases with increasing the applied current. Viceversa, class II neurons generate action potentials in a given frequency band and their frequency is quite insensitive to changes of the applied current.

In class I neurons the bifurcation of the equilibrium point (i.e. the quiescent state) is a saddle-node limit cycle bifurcation. This account for the integrate-and-fire generation of action potentials typical of these neurons. In this case the quiescent state disappears on a invariant limit cycle (i.e. the action potential).

In class II neurons the quiescent state loses its stability and a limit cycle appears through a Hopf bifurcation. In this case the neuron behavior is resonate-and-fire, since its response to a pair of stimuli depends on the timing between the stimuli relative so that it preferentially responds to given timings.

Another important difference is that the threshold in class II neurons is not well defined, whereas for class I neurons it is well defined. In fact, in class I the stable manifold of the saddle (i.e. the separatrix) separates two regions of the phase space having different qualitative behavior (i.e. relaxation to the quiescent state or action potential).

2.1.5 *Other neuron models*

In this section other neuron models derived by simplifying the Hodgkin-Huxley model are briefly reported. The first of these models is the Hindmarsh-Rose (HR) neuron which consists of a system of three first-order ordinary differential equations and is able to give a qualitative explanation of burst generation of action potentials [Hindmarsh (1984)]. This model is described by the following set of ordinary differential equations:

$$\begin{cases} \dot{x} = y + ax^2 - x^3 - z + I_{HR} \\ \dot{y} = 1 - bx^2 - y \\ \dot{z} = -rz + rS(x - c_x) \end{cases} \tag{2.8}$$

The model is intended to explain the generation of bursting activity in neuronal cells, which are initially silent, but once depolarized generate

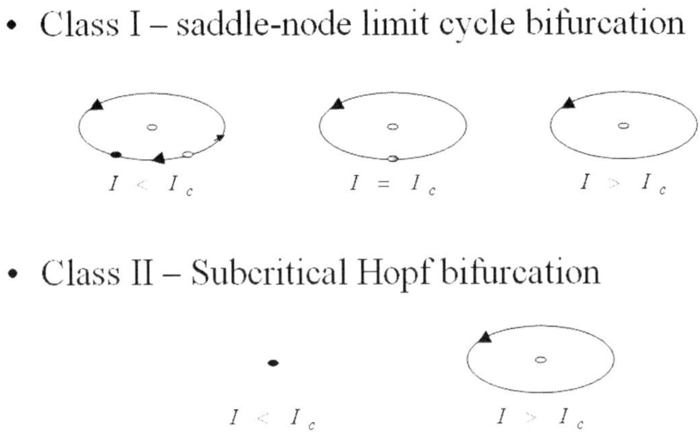

Fig. 2.5 Schematic view of bifurcations leading to class I and class II neuron excitability.

bursts of action potentials. For this reason Hindmarsh and Rose wrote the equations of a second-order system with three equilibrium points, one stable and two unstable. Moreover, the system has a stable limit cycle. The stable equilibrium point accounts for the silent state, while the stable limit cycle represents the action potential. The transition between the two stable states is triggered by a short pulse and represents the transition from the silent state to firing by a current pulse. They added a third equation to their original second-order model [Hindmarsh (1982)] to model slow adaptation of the discharge current. This equation aiming at modelling a slow adaptation phenomenon is characterized by a small parameter, $r = 0.0021$ in equations (2.8).

Another interesting model which in the following will be considered as elementary cell of a complex system of interacting units is the simplified model of the Inferior Olive (IO) neuron [Arena (2000); Giaquinta (2000)] described by the following equations:

$$\begin{cases} \dot{x} = \frac{1}{\varepsilon}[x(x-\gamma)(1-x) - y + I] \\ \dot{y} = -\Omega z + r(A - z^2 - r^2) \\ \dot{z} = \Omega r + z(A - z^2 - r^2) \end{cases} \tag{2.9}$$

where the quantity r is given as: $r = (\frac{y}{M} - x)$. The model mimics the salient features of a particular kind of neurons sited in the Inferior Olive region, a region of the brain sending sensory information to the cerebellum.

This model generalizes the FitzHugh model of neural excitability and considers as base state a limit cycle representing the subthreshold oscillations of the IO dynamics [Giaquinta (2000)]. In Equations (2.9) the variable x represents the cross membrane potential, while the variables y and z are recovery variables reproducing the dynamics of the most important ionic currents.

One neuron model, which conjugates the large variety of neuronal behaviors it can implement with a relatively low computational cost required to simulate it, is the model proposed by Izhikevich [Izhikevich (2003)]. It is described by the following equations:

$$\begin{aligned} \dot{v} &= 0.04v^2 + 5v + 140 - u + I \\ \dot{u} &= a(bv - u) \end{aligned} \tag{2.10}$$

with the spike-resetting

$$\texttt{if } v \geq 30\text{mV}\texttt{, then} \begin{cases} v \leftarrow c \\ u \leftarrow u + d \end{cases} \tag{2.11}$$

v and u are dimensionless variables representing neuron membrane potential and recovery variable, respectively. The time unit is ms. This model can reproduce the main firing patterns and neural dynamical properties of biological neurons such as spiking behaviors (tonic, phasic and chaotic spiking, class I and class II excitability) and bursting behavior, depending on the parameters chosen.

Many other models present very interesting properties. We refer to [Izhikevich (2000)] for a review.

2.2 Electronic Implementation of Neuron Models

There are many reasons for which an electronic implementation of neuron models may be desirable. Different motivations arise from the type of electronic implementation carried out. For instance, digitally-based architectures and VLSI systems offer the advantage of letting real-time simulations of many units, but on the other hand discrete component implementations of single cell dynamics (or of a few coupled neurons) require a faster design phase and may easily include tunable parameters, so that they can be used for instance to study small-size systems of neurons with respect to changes of the characteristic parameters.

In this section we will first discuss analog implementation of single cell dynamics and then a digital architecture for the simulation of many interacting units.

2.2.1 *Implementation of single cell neuron dynamics*

The single cell dynamics of a neuron model can be implemented by an electronic circuit by using different approaches. One neuron can be implemented by using a discrete component board and following for instance a design based on operational amplifiers and nonlinear devices. This approach is very effective, since it requires a short design phase which can be accomplished with the aid of CADs such as Spice, for instance. However, this approach does not allow the simulation of large-size systems.

Another possibility, which requires greater design efforts and higher costs, is to design the circuit at the transistor level and to integrate many units in VLSI.

Finally, a recent solution is the use of configurable analog circuits, which in some sense are an intermediate solution with respect to the VLSI and the discrete component approaches. In fact, the implementation based on configurable analog circuits requires a very fast design phase and many units (although the number is less than in the case of VLSI systems) can be implemented.

We adopted this approach to implement several neuron models. In particular, the device used for the implementation consists of a Field Programmable Analog Array (FPAA), the AN221E04 FPAA [Anadigm (2006)], which can be programmed by a software development tool called AnadigmDesigner2.

The core of the device is a two-by-two matrix of blocks named CAB (Configurable Analog Blocks) that can be connected with each other and with external I/O blocks. Each CAB contains a digital comparator, an analog comparator, two operational amplifiers and a set of capacitors. The programmability of FPAA systems is in fact provided by a the switched capacitor technology.

The CAB blocks are surrounded by the other subsystems. One section is dedicated to clock management, another to signal I/O and a digital section is devoted to IC configuration and dynamic reprogrammability. The digital section is based on a Configuration Logic State Machine mapping the interconnections inside the IC. This state machine allows the FPAA to be dynamically reprogrammed. A look-up-table provides the possibility of implementing a CAM generating arbitrary waveforms.

The number of CABs is four for each board, thus a single board can provide the possibility of implementing relatively simple dynamics. However, the Anadigm devices can be connected each other in a very simple way to build more complex circuits. One of these boards is connected to the serial port and allows the other boards to be programmed. The modularity of these boards allows to expand the number of CAMs available for its own design.

The software development tool allows the FPAA to be connected with the I/O ports and the desired circuit to be designed using pre-defined blocks. There is a large number of blocks called CAMs (Configurable Analog Modules) performing different functions. The main blocks useful to implement nonlinear circuits (and thus neuron dynamics) are: gain blocks, sum blocks, integrators, filters, multipliers, etc. All of them are provided by the development tool.

The internal operations of the FPAA are performed through voltage signals in differential mode. Thus, starting from the dimensionless equations of the dynamical system, an equivalent circuit is designed by assuming that the state variables are voltage signals.

The dynamic range of signals inside the FPAA device is bounded by physical constraints, so that in general the neuron state variables have to be scaled. Therefore, to implement neuron dynamics in the FPAA framework, one should: determine the dynamic range of the variables of the dimensionless equations and define new scaled variables according to the FPAA bounds; build a block scheme of the circuit, paying attention to the dynamic range bounds that have to be respected in all the blocks of the circuit; design the FPAA schematic by substituting idealized blocks with FPAA blocks; run the FPAA device and eventually adjust some parameters which may require some changes.

Fig. 2.6 shows oscilloscope traces referred to the FPAA implementation of several neuron dynamics. Fig. 2.6(a) refers to a circuit implementing the FHN neuron (2.7); both variables v and u are shown. Fig. 2.6(b) refers to a circuit implementing HR neuron dynamics (2.8). Fig. 2.6(c) refers to a circuit implementing the Izhikevich neuron model (2.10) with $a = 0.02$, $b = 0.2$, $c = -65$, $d = 6$ and $I = 0$. With such parameters the Izhikevich neuron and the circuit are characterized by regular spiking. Fig. 2.6(d) shows the state variables of a circuit implementing the Izhikevich neuron model (2.10) with $a = 0.02$, $b = 0.2$, $c = -50$, $d = 2$ and $I = 0$ so that tonic bursting emerges.

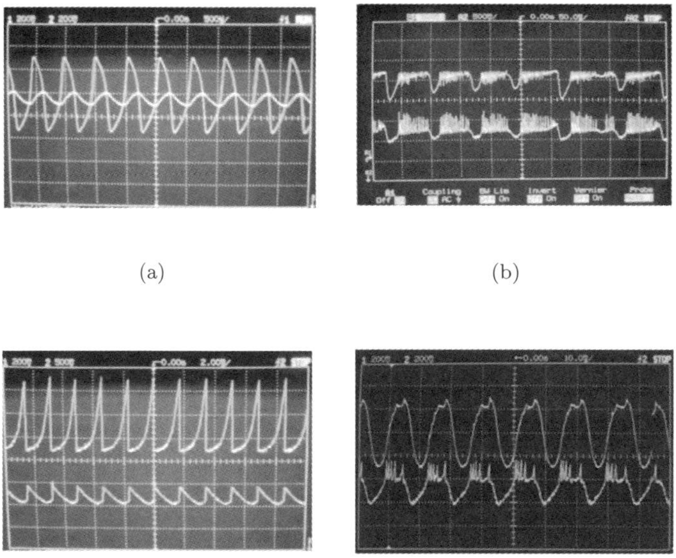

(a) (b)

(c) (d)

Fig. 2.6 FPAA implementation of neuron dynamics: (a) FHN neuron; (b) HR neuron; (c) Izhikevich neuron with regular spiking; (d) Izhikevich neuron with tonic bursting.

2.2.2 *Implementation of systems with many neurons*

When an electronic implementation of a system with many neurons is considered, the cost to implement the system (number of floating point operations to simulate one millisecond of the model) and the accuracy of the model become fundamental parameters for the choice of the model to be implemented. Therefore for a neural model it is necessary to have a good compromise between biological plausibility and implementation cost. The model introduced by Izhikevich satisfies these requirements.

In this section an electronic implementation of a system based on this model that emulates in efficient way the behaviour of cortical neurons is discussed. First of all, an analysis of feasibility has been carried out in order to verify the possible implementation of the Izhikevich's model on digital hardware, more precisely on FPGA (Field Programmable Gate Array). The benefits of the use of this device are connected with its reprogrammability, in fact this device is extremely versatile in the implementation of different functional specifics. In addition FPGA permits to implement real-time systems such as neural models, to work with adjustable frequencies in the range of megahertz and to integrate many components (neural cells) on the same chip. Thanks to these characteristics it is possible to implement on FPGA complex systems such as networks of neurons. The high configurability of the device allows the user to analyze the behavior of the implemented system with respect to different parameters and inputs. Moreover, the output of the device can be simply manipulated, through calculators or logic states analyzers, and mostly important can be interfaced with external systems as biological ones.

The first step needed for the FPGA-based implementation of the Izhikevich's neuron is a fixed point simulation of this system using a fixed number of bits to represent membrane potential and recovery variables. An analysis of feasibility has been done to investigate if this neural model in spite of quantization continues to give biological coherent results. The operation of the quantization is functional for the hardware implementation. A discrete-time version of the Izhikevich's model, obtained by using the Euler integration method, is used. After a characterization of the required accuracy, it was found that twentyfour bits are needed to represent the samples of the two state variables, eight of these bits are needed for the integer part and sixteen for the decimal part. This choice is connected with the range $[-70mV, +30mV]$ in which the membrane potential variable varies. The accurate precision of the decimal part is connected with the operations (sums and products) carried out during each iteration for the computation of the next samples. The results of the comparison between neural dynamics obtained through numerical simulations in floating point and those in fixed point are satisfying.

The FPGA implementation of the quantized model of Izhikevich's neuron is based on a flow of planning which can be divided in three steps: source code implementation (VHDL / VERILOG); synthesis; place and route.

VHDL (Hardware Description Language) is a language used for the implementation of particular functions in digital hardware. The source code is written by using a descriptive approach. In the synthesis phase the VHDL description is converted into a scheme based on logic gates. Place and route consists of the placement and physical layout of the logic gates determined in the synthesis phase, into the portion of silicon available, in way to obtain the desirable hardware. This phase is also required to solve hardware specifics such as timing, area etc. connected with the project and therefore with the components of the structure implemented. For the implementation of the Izhikevich's neuron on FPGA a structure based on different functional modules was used. Each module works together with the others modules to emulate the behavior of cortical neurons.

The schematic in Fig. 2.7 shows the constitutive modules of the system implemented on FPGA. The first module is the interface module. This module is necessary to interface the system with the external world. In particular, the signals sent by the PC are processed by this module. This signal and the others shown in Fig. 2.7 are used to program the system. For instance, the signal 'Type' allows the user to select the kind of neural dynamic to process (tonic spiking, tonic bursting, phasic spiking and so on). Another signal needed by the interface module is the clock signal. The possibility to tune its value enables the system to operate with a velocity comparable or higher to that of the biological model. In other words, if needed the same information processing operated by real cortical neurons can be obtained in a shorter time. The system shown in Fig. 2.7 is also provided of a ROM (Read Only Memory) with the aim of storing the sets of parameters necessaries to obtain the main dynamics of cortical neurons. It is important to underline that among the parameters stored in the ROM the input current I is not included, in fact this parameter is directly sent as input from the outside. In this way with the assistance of a Pattern Generator it is possible to send any arbitrary (either excitatory or inhibitory) waveform as input of the system in order to investigate the system response to arbitrary input. This makes the system extremely versatile and a powerful tool to study the behavior of cortical neurons. The centre of coordination of the modules implemented on FPGA is called FSM (Finite State Machine), this is a component that is interfaced with the memory and the module 'Neuron', analyzed below. FSM is a combinatory-sequential net that passes from a state to the next in relation to the event that is verified.

The module called 'Neuron' is the functional core of the implemented system. In fact it is devoted to the calculation of the samples of v and u and of the reset mechanism. This module is based on two fundamental functional entities. The first is purely combinatory and manages the resolution of the equations to compute the samples of the two state variables. The second block manages the reset condition, therefore controlling the samples at every time; moreover, it resets the two variables when the resetting condition is met. At this point the second constitutive entity supplies the update values again to the first block for the computation of the new sam-

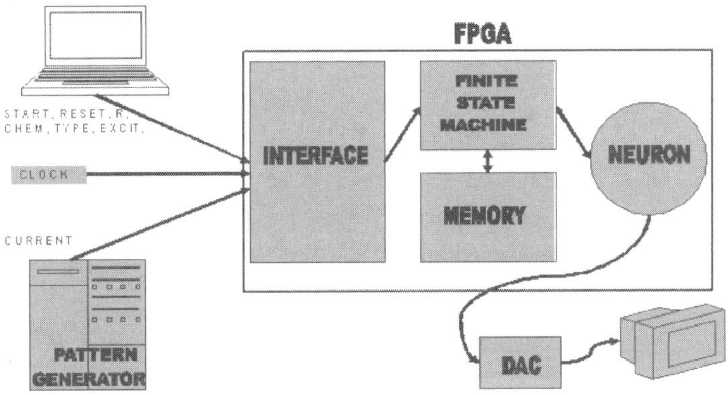

Fig. 2.7 Schematic of the FPGA-based implementation of Izhikevich's neuron model.

ples. Once defined the architecture for the single neuron implementation, simple networks of interconnected neurons have been implemented with the aim of exploring the capabilities of the system to implement many neural cells on the same chip. The purpose of the system is in fact to emulate the complex behavior of biological neural network.

A structure of two connected neurons was implemented, in this case the output (i.e. the membrane potential) of one neuron becomes fundamental to compute the output of the other neuron, thus the general structure of the architecture of FPGA shown in Fig. 2.7 has been properly changed by taking into account another 'Neuron' module and the interface between this block and the other neuron. The approach can be extended to the more general case of n neurons. As in the case of one neuron implementation, some external signals are sent from the outside to make the system fully adaptive and able to implement any neural behavior that has to be tested. Through these signals, the input current waveforms supplied to each cell can be independently changed.

Fig. 2.8 shows two examples of the different neural dynamics studied through our FPGA system. The waveforms obtained are the response of the system to a current step given as input to the FPGA neuron. The results obtained agree with the software simulations. In this case the clock frequency of the FPGA was fixed to 1MHz. This frequency (which can be arbitrarily varied) permits to control the typical time scales of the implemented neuron and thus allows us to obtain neural system faster than biological ones. This is particularly important since numerical simulations of large arrays of neurons may be time-consuming.

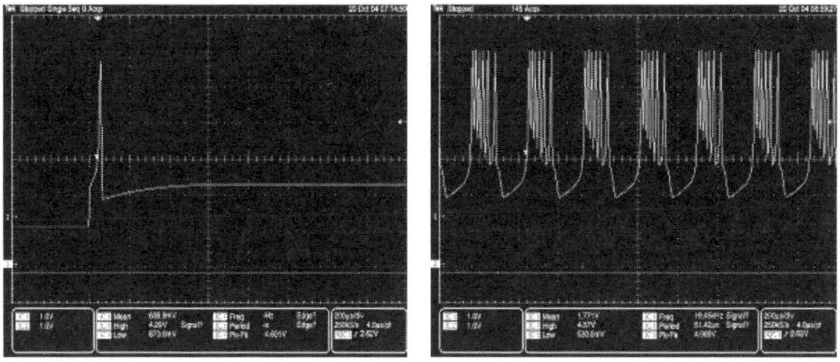

(a) (b)

Fig. 2.8 Neural dynamics of cortical neurons implemented on FPGA (left: phasic spiking, right: tonic bursting).

In Fig. 2.9 an example of the behavior exhibited by a digital system of two coupled neurons is shown. The two neurons have different dynamics (one is characterized by tonic spiking, the other by tonic bursting). The synapse was fixed to be chemical, the two neurons exhibit out-of-phase synchronization.

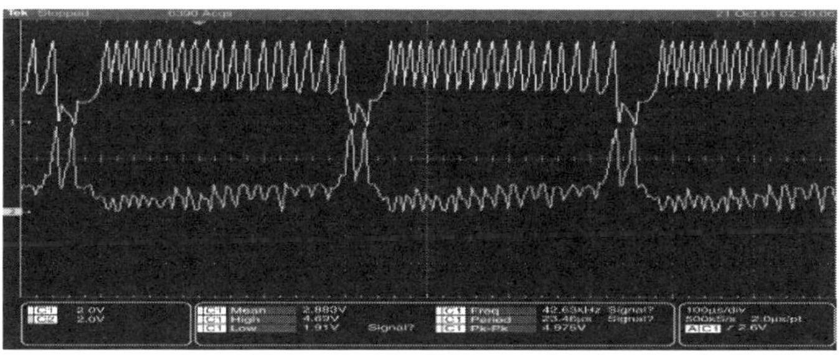

Fig. 2.9 Two cortical neurons implemented on FPGA, connected by a bidirectional excitatory chemical synapse and having different dynamics (upper trace: tonic bursting, lower trace: tonic spiking).

2.3 Local Activity Theory for Systems of IO Neurons

Until now neurons have been considered only under the viewpoint of single cell dynamics. In this and in the following sections we take into account many neurons connected to form a complex system. The importance of the study of complex systems is witnessed by the large number of fields and contexts in which the term complexity is used: biology, architecture, computer science, engineering, physics. Complex systems show organization without a central organizing unity. Their peculiar characteristics are adaptation, self-organization and emergence. Although "engineering complex systems" [Ottino (2004)] is a process still to be investigated, ideas such as including self-organization as a part of the design are appealing for the scientific community of circuits and systems engineers.

Several scientists have analyzed complex systems and formulated very important concepts such as order from disorder [Schrodinger (1948)], self-organization [Nicolis (1997)], cooperation and synergetics [Haken (1983)], edge of chaos [Langton (1990)]. These concepts are now key points of the theory of complexity.

Moreover, recent developments have lead to new approach based on analytical tests [Chua (1997)]: the local activity theory and the Edge of Chaos theorem allow us to find analytical criteria to determine the regions in which emergent patterns may arise. These unique analytical tools based on circuit theory that allow one to investigate on complex systems have been applied to many dynamical systems [Chua (1989); Dogaru (2003)]. In this section we apply the theory of local activity to the case of a complex systems made of IO neurons. The criteria derived from this theory allow one to plot bifurcation diagrams illustrating the local activity regions for this system and thus to draw important conclusions on the study of networks of IO neurons. Moreover, the results discussed in the following lead to the conclusion that Reaction Diffusion Cellular Nonlinear Networks (RD-CNN), based on array of locally active devices (the IO neurons), represent a new wave-based computational paradigm for image processing. Furthermore, it is shown how the 3D Reaction-Diffusion (RD) generalized CNN, introduced in [Arena (2005)], allows us to process 3D shapes as does occur for 2D images.

2.3.1 *The theory of local activity for one-port and two-port systems*

The theory of local activity gives analytical criteria to establish whether a regular grid of dynamical systems may exhibit emergent behavior [Chua (1997)]. As in the following the local activity criteria will be applied to both one-port and two-port third-order systems, in this section we briefly recall the passivity criteria for a one-port and two-port systems.

Definition. A linear time-invariant one-port system is locally passive at an equilibrium point Q, if, and only if, its cell impedance $Z_Q(s)$ at Q satisfies the following conditions.

- P1. $Z_Q(s)$ has no pole in $Re(s) > 0$.
- P2. $Re[Z_Q(j\omega)] \geq 0 \; \forall \omega \in \mathbb{R}$ where $s = j\omega$ is not a pole of $Z_Q(s)$.
- P3. If $s = s_p$ is a pole of $Z_Q(s)$ on the imaginary axis, i.e. $s_p = j\omega_p$, then s_p must be a simple pole and its residue must be a positive real number.

The locally active region is given by the complementary set of the locally passive region. The edge of chaos region is given by the locally active region, where the P1 condition is satisfied.

In the case of nonlinear systems the previous conditions have to be applied to the linearized system for each equilibrium point. It has been proved [Chua (1997)] that the local activity represents a *sine qua non* condition, for which a regular lattice cannot generate patterns.

Definition. A linear time-invariant two-port system is locally passive{Passivity} at an equilibrium point Q, if, and only if, its cell impedance matrix $\mathbf{Z}_Q(s)$ at Q satisfies the following conditions.

- P1. $\mathbf{Z}_Q(s)$ has no pole in $Re(s) > 0$.
- P2. $\mathbf{Z}_Q^H(j\omega) \triangleq \mathbf{Z}_Q^{+}(j\omega) + \mathbf{Z}_Q(j\omega)$ is positive semi-definite $\forall \omega \in \mathbb{R}$ where $s = j\omega$ is not a pole of $\mathbf{Z}_Q(s)$.
- P3. If $s = s_p$ is a pole of $Z_Q(s)$ on the imaginary axis, i.e. $s_p = j\omega_p$, then s_p must be a simple pole and its residue matrix

$$\mathbf{K}_{-1} = \begin{cases} \lim_{s \to j\omega_p} (s - j\omega_p)\mathbf{Z}_Q(s) & \texttt{if } \omega_p < \infty \\ \lim_{\omega_p \to \infty} \frac{\mathbf{Z}_Q(s)}{j\omega_p} & \texttt{if } \omega_p = \infty \end{cases}$$

is a positive semi-definite Hermitian matrix.

As in the case of one-port system the locally active region is given by the complementary set of the locally passive region. The edge of chaos region is given by the locally active region, where the P1 condition is satisfied.

2.3.2 *The local activity and the edge of chaos regions of the inferior olive neuron*

We assume that the IO model is either a one-port or a two-port system, where inputs I_1 and I_2 act on the dynamic equation of variable x or y, respectively. We still maintain the term I to account for a current term specific of each IO neuron. Restricted local activity criteria (i.e. $I_1 = I_2 = 0$) have been applied. The IO model has been analyzed with respect to all the parameters characterizing it, however in the following we only focus on the parameters which have been revealed to be the most important (γ, Ω and I), while keeping fixed the others ($M = 0.5$, $\varepsilon = 0.01$ and $A = 0.005$).

The first step to apply the local activity theory is to calculate the equilibrium points for the system represented by the set of equations (2.9). By letting $\dot{x} = \dot{y} = \dot{z} = 0$, one obtains:

$$\begin{cases} x(x-\gamma)(1-x) - y + I = 0 \\ -\Omega z + r(A - z^2 - r^2) = 0 \\ \Omega r + z(A - z^2 - r^2) = 0 \end{cases} \tag{2.12}$$

The nonlinear algebraic system (2.12) can be rewritten in terms of an unique nonlinear characteristic equation as follows:

$$x^3 - (1+\gamma)x^2 + (\gamma + M)x - I = 0 \tag{2.13}$$

which gives the set $\bar{x}$ of the equilibrium points. The other two coordinates can be found by:

$$\begin{cases} \bar{y} = M\bar{x} \\ \bar{z} = 0 \end{cases} \tag{2.14}$$

The equation (2.13) admits one or three solutions, so that system (2.9) has one or three equilibrium points. Once found the equilibrium points Q_i, the system has been linearized around them and the corresponding transfer function has been calculated.

The Jacobian $A = J_Q$ of the system is given by:

Table 2.1 Legend of parameter regions.

	Number of real solutions	Number of stable equilibriums	Region
	3	3	Locally Passive
	1	1	Locally Passive
	3	> 1	Edge of Chaos
	1	1	Edge of Chaos
	3	0	Locally Active at all equilibriums
	1	0	Locally Active at the equilibrium
	3	≤ 2	Locally Active in at least one equilibrium

$$A = J_Q = \begin{pmatrix} -\frac{3}{\varepsilon}\bar{x}^2 + \frac{2+2\gamma}{\varepsilon}\bar{x} - \frac{\gamma}{\varepsilon} & -\frac{1}{\varepsilon} & 0 \\ -A & -\frac{A}{M} & -\Omega \\ -\Omega & \frac{\Omega}{M} & A \end{pmatrix} \tag{2.15}$$

Once obtained the Jacobian matrix A, either the one-port or two-port criteria for local activity can be applied as discussed in details in [Min (2000)] for one-port third-order cells and in [Yang (2001)] for the two-port case, respectively. This allows one to obtain the edge of chaos, local activity and passivity regions for the IO model.

As concerns the two-port case, the complete bifurcation diagram related to the parameters Ω, γ and I is reported in Fig. 2.10. The regions are coded by using the color coding adopted in [Dogaru (2003)] and detailed in Table 2.1. Light and dark colors refer to the case of one or three equilibrium points, respectively. The range of the admissible parameter values is considered. Moreover, the previous map can be split in several magnified regions as reported in Figs. 2.11 and 2.12, respectively.

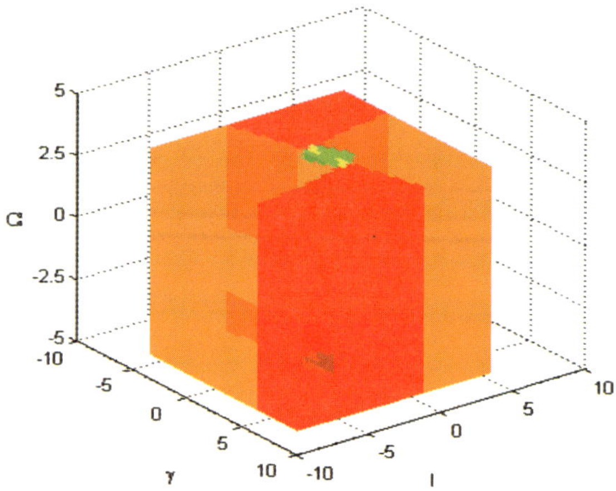

Fig. 2.10 Bifurcation diagram of the IO neuron with respect to the parameters I, γ and Ω for the two-port case.

2.4 Simulation of IO Systems: Emerging Results

Numerical simulations show that in many cases, when the parameters of the model are chosen in locally active regions, emergent phenomena such as autowaves and spiral waves appear. In this section we will discuss in some details the results related to the emergent capabilities of image processing shown by the IO complex system.

2.4.1 *The paradigm of local active wave computation for image processing*

Let us consider a classical problem of image processing: the contour detection of a planar image. Let us consider a 2D lattice of RD-CNN of 60×60 IO neurons as follows:

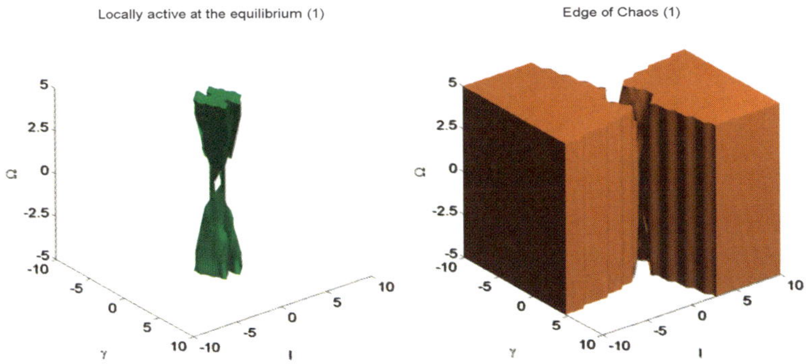

(a) (b)

Fig. 2.11 Bifurcation diagram of the IO neuron with respect to the parameters I, γ and Ω for the two-port case. (a) Magnification of the locally active region at one equilibrium. (b) Magnification of the Edge of Chaos region at one equilibrium.

$$\begin{cases} \dot{x}_{i,j} = \frac{1}{\varepsilon}[x_{i,j}(x_{i,j} - \gamma)(1 - x_{i,j}) - y_{i,j} + I + D1 * \nabla x_{i,j}] \\ \dot{y}_{i,j} = -\Omega z_{i,j} + r_{i,j}(A - z_{i,j}^2 - r_{i,j}^2) + D2 * \nabla y_{i,j} \\ \dot{z}_{i,j} = \Omega r_{i,j} + z(i,j)(A - z_{i,j}^2 - r_{i,j}^2) \end{cases} \tag{2.16}$$

with $r_{i,j} = (\frac{y_{i,j}}{M} - x_{i,j})$, $1 \leq i \leq 60$, $1 \leq j \leq 60$ and

$$\nabla x_{i,j} = x_{i-1,j} + x_{i,j-1} + x_{i,j+1} + x_{i+1,j} - 4x_{i,j}$$

$$\nabla y_{i,j} = y_{i-1,j} + y_{i,j-1} + y_{i,j+1} + y_{i+1,j} - 4y_{i,j}$$

Let us consider a RD-CNN (2.16) where each cell is locally active. In particular, the parameters have been chosen as follows: $I = 0.8$, $\gamma = 3$ and $\Omega = -9$, which correspond to the region shown in Fig. 2.12(b). Boundary conditions are zero-flux; the assumed diffusion coefficients are $D_1 = 1$ and $D_2 = 15$.

The performed experiment is referred to the contour detection of a given planar image. The original image to be processed is given as initial condition of the variable x of the cells in the lattice that characterizes the image to be processed. In Fig. 2.13 the results are reported. It is evident how the RD-CNN is able to process the image to extract the contour in short time.

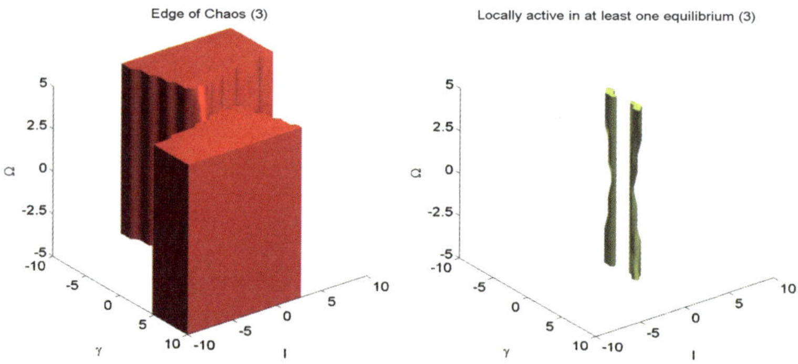

(a) (b)

Fig. 2.12 Bifurcation diagram of the IO neuron with respect to the parameters I, γ and Ω for the two-port case. (a) Magnification of the Edge of Chaos region at three equilibriums. (b) Magnification of the locally active region, where at least one of the three equilibriums is unstable.

A key point of this approach is that the contour detection, unlike the image processing performed by other CNN algorithms [Chua (2002)], occurs thanks to the reaction-diffusion process. By changing the reactive part of the reaction-diffusion equation, according to the local activity theory, different emerging phenomena arise.

2.4.2 *Local active wave computation based paradigm: 3D-shape processing*

In this section the experiments related to a 3D RD-CNN lattice are considered. The equations of the 3D RD-CNN are the following:

$$\begin{cases} \dot{x}_{i,j,k} = \frac{1}{\varepsilon}[x_{i,j,k}(x_{i,j,k}-\gamma)(1-x_{i,j,k}) - y_{i,j,k} + I + D_1\nabla x_{i,j,k}] \\ \dot{y}_{i,j,k} = -\Omega z_{i,j,k} + r_{i,j,k}(A - z_{i,j,k}^2 - r_{i,j,k}^2) + D_2\nabla y_{i,j,k} \\ \dot{z}_{i,j,k} = \Omega r_{i,j,k} + z_{i,j,k}(A - z_{i,j,k}^2 - r_{i,j,k}^2) \end{cases} \tag{2.17}$$

where $r_{i,j,k} = (\frac{y_{i,j,k}}{M} - x_{i,j,k})$, $1 \le i \le 40$, $1 \le j \le 40$, $1 \le k \le 40$ and

$$\nabla x_{i,j,k} = x_{i-1,j,k} + x_{i,j-1,k} + x_{i,j,k-1} + x_{i+1,j,k} + x_{i,j+1,k} + x_{i,j,k+1} - 6x_{i,j,k}$$

$$\nabla y_{i,j,k} = y_{i-1,j,k} + y_{i,j-1,k} + y_{i,j,k-1} + y_{i+1,j,k} + y_{i,j+1,k} + y_{i,j,k+1} - 6y_{i,j,k}$$

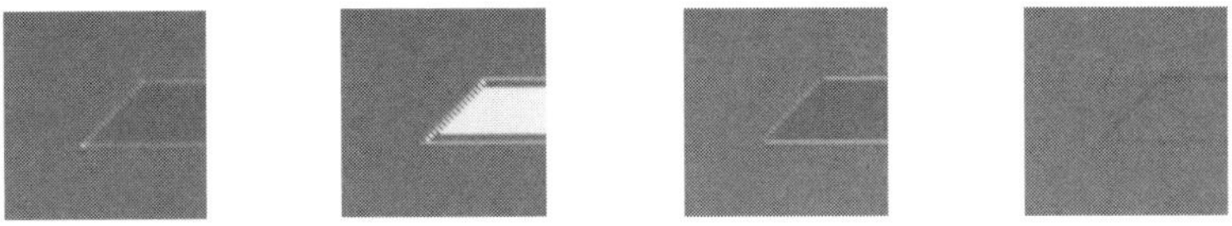

(a) (b) (c) (d)

Fig. 2.13 Contour detection of 2D images by a RD-CNN of locally active (in at least one equilibrium) cells. Several snapshots of the dynamic evolution of the RD-CNN at different times are shown. Each pixel in each snapshot represents the color-coded value of variable $x_{i,j}(t)$ (the maximum value $x_M \simeq 2.9$ and the minimum value $x_m \simeq 0$ correspond to red and blue, respectively). (a) $t = 0.4$ s. (b) $t = 0.6$ s. (c) $t = 0.7$ s. (d) $t = 10.0$ s.

The parameters have been fixed as follows: $I = 0.8$, $\gamma = 3$ and $\Omega = -9$, which correspond to the local activity region. Zero-flux boundary conditions and diffusion coefficients $D_1 = 1$ and $D_2 = 15$ have been chosen. This RD-CNN represents the 3D generalization of the RD-CNN (2.16), which was demonstrated able to extract the contour of planar images. The 3D image processing capabilities of the RD-CNN (2.19) to perform the contour detection of a 3D shape are evident in Figs. 2.14 and 2.15: given an initial cube its six sides are extracted. Figure 2.14 shows how starting from a filled cube, after the processing the inner and outer surfaces are visible. The cube is obtained by selecting initial conditions as follows:

$$x_{i,j,k}(0) = y_{i,j,k}(0) = z_{i,j,k}(0) = \begin{cases} 1 \texttt{ if } 20 < i,j,k < 30 \\ -1 \texttt{ otherwise} \end{cases} \tag{2.18}$$

Figure 2.15 shows three sections of the processed shape, confirming that only the sides of the initial cube have been extracted: starting from a filled cube of cells which have a red value (i.e. $x_{ijk} \simeq 2.9$), at the end of the contour extraction only the cells belonging to the six sides of the cube have a steady-state red value.

Moreover, another experiment related to edge detection has been considered. In this case a slightly different model with two further self-feedback terms has been considered as follows:

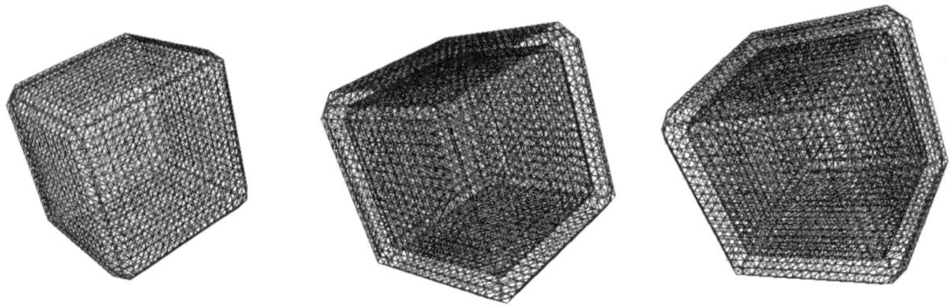

(a) (b) (c)

Fig. 2.14 Contour detection of a 3D shape by a RD-CNN of locally active IO neurons. Several snapshots of the dynamic evolution of the RD-CNN are shown. Only cells with $x_{i,j,k} \simeq 2.9$ are shown. (a) $t = 4$ s. (b) $t = 10$ s. (c) $t = 15$ s.

(a) (b) (c)

Fig. 2.15 Contour detection of a 3D shape by a RD-CNN of locally active IO neurons. The adopted color scale is normalized to the maximum value ($x_M \simeq 2.9$, red) and the minimum value ($x_m \simeq 0$, blue). Several sections of the steady-state shape shown in Fig. 2.14 are reported: (a) $i = 20$; (b) $j = 20$; (c) $k = 20$.

$$\begin{cases} \dot{x}_{i,j,k} = \frac{1}{\varepsilon}[x_{i,j,k}(x_{i,j,k} - \gamma)(1 - x_{i,j,k}) - y_{i,j,k} + 2x_{i,j,k} + I + D_1 \nabla x_{i,j,k}] \\ \dot{y}_{i,j,k} = -\Omega z_{i,j,k} + r_{i,j,k}(A - z_{i,j,k}^2 - r_{i,j,k}^2) + 2y_{i,j,k} + D_2 \nabla y_{i,j,k} \\ \dot{z}_{i,j,k} = \Omega r_{i,j,k} + z_{i,j,k}(A - z_{i,j,k}^2 - r_{i,j,k}^2) \end{cases} \quad (2.19)$$

The other parameters have been chosen as in the previous case. In Figs. 2.16 and 2.17 the related experiments are shown. In Fig. 2.16 starting from a filled cube the RD-CNN is able to extract the edges. Another experiment with a more complex shape is shown in Fig. 2.17. The experiments reported above show how RD-CNNs of neuron cells can process 3D shapes.

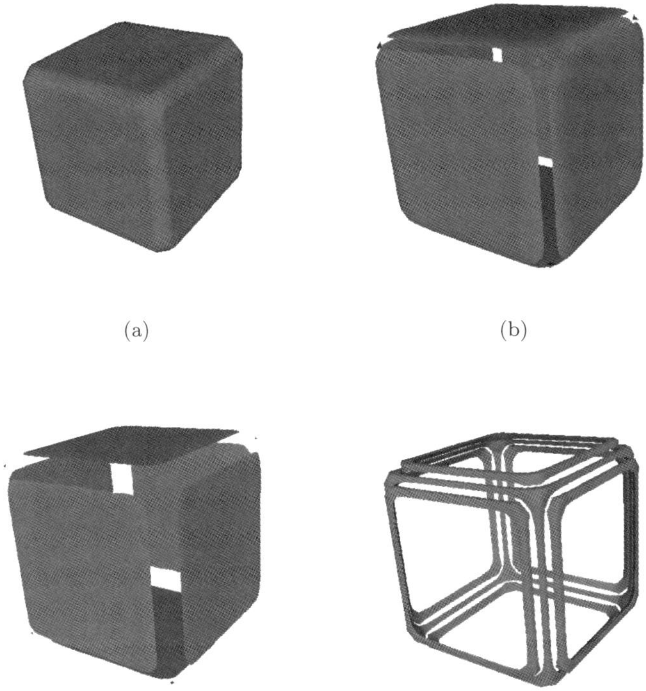

(a) (b) (c) (d)

Fig. 2.16 Edge detection of 3D shapes by a locally active (in at least one equilibrium point) RD-CNN: first example. Only cells with $x_{i,j,k} \simeq 2.5$ are shown. Several snapshots of the dynamic evolution are shown: (a) $t = 4$ s; (b) $t = 11$ s; (c) $t = 12$ s; (d) $t = 100$ s.

2.5 Networks of HR Neurons

In this section we investigate the role of topology on synchronization of networks of HR neurons, a fundamental feature of many technological and biological fields. We study it in HR neural networks, with electrical and chemical synapses, where neurons are placed on a bi-dimensional lattice, folded on a torus, and the synapses are set according to several topologies.

The analyzed topologies belong to two different classes: the static one (where the number of nodes and edges is constant in time) and the dynamical one (where at each time step a new node and some edges are added).

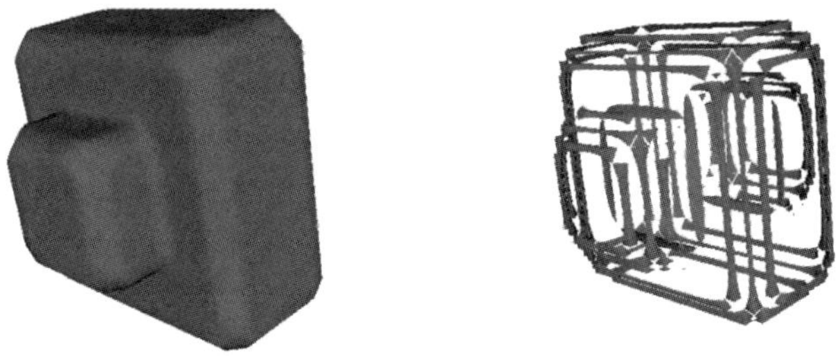

(a) (b)

Fig. 2.17 Edge detection of 3D shapes: second example. (a) Initial shape. (b) $t = 24$ s (steady-state shape).

Different neural networks, characterized by different topologies and different synapses, are compared and characterized by means of both static parameters, that take into account only the structure, and dynamic parameters, that take into account the temporal evolution of the state variables of neurons. We consider nodes placed on a bi-dimensional surface with periodic boundary conditions so that the surface is a torus and edge effects are avoided. In particular, the following topologies are analyzed: regular lattices; Erdös-Rényi (ER) random graphs [Erdös (1959)]; the Watts and Strogatz model of small-world networks (WS); the Newman and Watts model of small-world networks (NW); the Barabási and Albert [Barabási (1999)] model of scale-free network (BA) and generalized Barabási-Albert Scale-free model (GBA) introduced in [Cosenza (2005)].

Among the parameters shown to analyze the networks, it has been also included a cost parameter with a two-fold aim: on one side it allows us to take into account the role that physical distance plays in the formation of new synapses (edges) between neurons (nodes); on the other it can be a fundamental parameter for the design of artificial neuron networks.

2.5.1 *The neural model*

The networks analyzed consists of coupled HR units. The parameters values have been chosen in order to obtain a chaotic behavior ($a = 3$, $I = 3.281$, $b = 5$, $r = 0.0021$, $s = 4$, $c_x = 1.6$).

Two different types of synapses are considered: the electrical one and the chemical one. In both cases the synaptic current I_{syn} is added to the right-side of the first differential equation of Eq. (2.8).

The electrical synapse is based on the diffusion process, and is modelled as a resistance. For a particular node j the synaptic current depends on the diffusion coefficient and on the membrane potential of itself and of its first neighbors:

$$I_{syn,j}(t) = D \cdot \sum_{i \in \mathbf{G}} a_{ij} \cdot (x_i(t) - x_j(t)), \tag{2.20}$$

where D is the diffusion coefficient. In this kind of synapse the delay is negligible and the electrical coupling results in levelling of the membrane potential.

The chemical synapse is the place where electrical signals coming from the axon of the pre-synaptic neuron are transduced in chemical signals (in the synaptic cleft), and then again in electrical signals to travel in the dendrite of the post synaptic neuron. Such a process can take place thanks to particular chemical intermediaries, called neurotransmitters. When a presynaptic event occurs, a certain quantity of neurotransmitters is released in the synaptic cleft. The neurotransmitters spread quickly in the synaptic channel towards the post-synaptic site, where form a chemical bond with the post-synaptic receptors. This gives rise to the increase of the conductance of the synapse, and hence to the presence of a post-synaptic current. The post-synaptic current I_{syn} is, therefore, directly proportional to the percentage of receptors bound with neurotransmitters [Bove (1998)]:

$$I_{syn,pre \rightarrow post}(t) = g_{syn,pre \rightarrow post}(t) \cdot [E_{syn} - x_{post}(t)] \tag{2.21}$$

$$g_{syn,pre \rightarrow post}(t) = g_{synMax} \cdot r_{pre \rightarrow post}(t) \tag{2.22}$$

$$r_{pre \rightarrow post}(t) = \begin{cases} [r_{pre \rightarrow post}(t_0) - r_\infty] \cdot e^{(t-t_0)/\tau_r} + r_\infty & if\ t_0 < t \leqslant t_0 + \tau \\ r_{pre \rightarrow post}(t_0 + \tau) \cdot e^{-\beta \cdot (t - t_0 - \tau)} & if\ t > t_0 + \tau \end{cases} \tag{2.23}$$

$$r_\infty = \frac{\alpha \cdot T_{max}}{\alpha \cdot T_{max} + \beta} \tag{2.24}$$

$$\tau_r = \frac{1}{\alpha \cdot T_{max} + \beta} \quad , \tag{2.25}$$

where $g_{syn}(t)$ is the actual conductance of the synapse at time t, g_{synMax} is the maximum conductance of the synapse (i.e. when the totality of post-synaptic receptors are bound with neurotransmitters), $r(t)$ is the percentage of receptors bound with neurotransmitters, t_0 is the time at which the pre-synaptic event occurs, τ_r is a characteristic time constant, $T_{max} = 1$ is the maximum concentration of neurotransmitters at time t_0, α, β and τ are parameters that we have fixed respectively to $\alpha = 2$, $\beta = 1$ and $\tau = 2$, and E_{syn} is the synaptic reversal potential (equal to 1.52 for an excitatory synapse and -1.36 for an inhibitory synapse), i.e. the value of the membrane potential for which no synaptic current flows in the ionic channel. Due to an electrical-chemical transduction, the chemical synapse introduces a delay, that is not constant for real neural networks [Abarbanel (1996)]. The chemical synapse, differently from the electrical one, is directed, in the sense that exists a preferential direction in the information flux. Hence, if communication between nodes i and j can occur in both directions, it is due to the presence of two different uni-directional synapses, each with its parameters and its actual conductance.

2.5.2 *Parameters for dynamical analysis*

In this paragraph some parameters, that will be useful for comparing performance of each network, are introduced.

The *structural cost* of a network G is the total physical length of the edges present in G defined as follows:

$$Cost(\mathbf{G}) = \sum_{i,j \in \mathbf{G}} a_{ij} \cdot l_{ij}, \tag{2.26}$$

where a_{ij} is the element ij of the adjacency matrix and l_{ij} is the Euclidean distance between nodes i and j.

In order to evaluate the degree of synchronization of a system, made of coupled units, we define the *synchronization index* σ [Fortuna (2004)]. Let matrix A collect in its rows all the N signals generated by the N subunits of the system and let $C^{(NxN)} = A \cdot A^T$ be the covariance matrix. The synchronization index takes into account the eigenvalues of C (i.e. the squares of the singular values of the matrix A). If all the signals are uncorrelated, all the singular values will be non-null. If all the signals are identical we will find that only one singular value is different from 0, i.e. the rank of the matrix C is equal to 1. Of course if signals are similar, but not identical, we will find very small but not null singular values. The synchronization index is, thus, defined at a certain percentage ξ, as the minimum number m of eigenvalues, whose sum is greater than a percentage of the trace of C:

$$\sigma(\xi) = \min\ m \ \mid \sum_{i=1}^{m} \lambda_i^{sort} > \xi \cdot Tr(C), \tag{2.27}$$

where λ_i^{sort} is the i-th largest eigenvalue of the Covariance matrix C.

If we have N neurons (and then N signals), this index varies in the range from 1 (when all signals are synchronized) to $[\xi N] + 1$ (when there is not synchronization) and gives information about the total number of different dynamics present in the system. In the following we will consider the synchronization index to the 95%, because $N = 196$, $1 \leq \sigma(0.95) \leq 187$.

Another important parameter here introduced is the *average power consumption* in information exchange. In analogy with electrical systems, the average power consumption is defined in two different ways, depending on the particular synapse model used. As far as the case with electrical synapses is concerned, we define the average power consumption $P_m(t_0, t_1)$ from time t_0 to time t_1 as:

$$P_m(t_0, t_1) = \sum_{i \neq j \in G} \frac{1}{t_1 - t_0} \cdot \int_{t_0}^{t_1} a_{ij} \cdot D \cdot \frac{1}{l_{ij}} \cdot [x_i(t) - x_j(t)]^2 dt \tag{2.28}$$

with

$$\frac{1}{R_{ij}} = a_{ij} \cdot D \cdot \frac{1}{l_{ij}} \tag{2.29}$$

where R_{ij} represents the resistance between nodes i and j, a_{ij} is the element ij of the adjacency matrix and l_{ij} is the physical distance between nodes i and j.

The average power consumption for the electrical synapses strictly depends on the differences of potentials $x_i - x_j$. For instance, if from time t_n the system is perfectly synchronized, i.e. $x_i(t) = x_j(t)$ for each i and j and for $t > t_n$, the average power consumption $P_m(t_n, +\infty)$ is null.

The average power consumption for the case of chemical synapses is instead defined as:

$$P_m(t_0, t_1) = \sum_{i \neq j \in G} \frac{1}{t_1 - t_0} \cdot \int_{t_0}^{t_1} a_{ij} \cdot g_{syn,ij}(t) \cdot [E_{syn} - x_j(t)]^2 dt, \tag{2.30}$$

where $g_{syn,ij}$ is the synaptic conductance between nodes i and j and E_{syn} is the synaptic potential. The average power consumption for chemical synapses, here defined, considers only a localized dissipation, that does not takes into account the length of axons and dendrites.

2.5.3 *Dynamical effects of topology on synchronization*

In this section simulation results related to both networks with electrical synapses and chemical synapses are discussed. In both cases we have considered networks with $N = 196$ nodes.

First, we investigated static topologies with fixed average degree $< k >= 4$ as a function of the diffusion coefficient D. We have chosen the average degree $< k >= 4$, because in such a way we are able to distinguish more clearly the effects of different topologies. With greater values of $< k >$, instead, in all the models the number of edges increases, leading towards a fully connected network and, thus, hampering a clear differentiation of the effects of topologies.

The regular lattice, thanks to its good local properties, reaches a minimum value $\sigma = 3$ for a diffusion coefficient $D = 0.9$. It means that the system has not reached a perfect synchronization, but, nevertheless, it shows a good synchronization for low frequency behavior (bursts are simultaneous, spikes are asynchronous). The ER graph also reaches very low values of the synchronization index σ. It means that in a network with electrical synapses, global properties weight more than local ones. Nevertheless, the ER graph costs more than five times the cost of the lattice.

Even a better synchronization is reached by the WS model. It demonstrates that, though global properties weight more than local ones, they are both of fundamental importance for synchronization. Moreover, the WS model (for the chosen value of $p = 0.08$) has a structural cost much lower than the ER graph. The WS model is an optimal choice, as regards synchronization with the constrain of low building cost.

The analysis discussed above is centered on a fixed value for the average degree ($< k >= 4$). Now we take into consideration the NW model, for which $< k >= 4$ is only the initial condition, and the average degree grows with the probability q of adding new edges.

A comparison of static and dynamical models is presented in Fig. 2.18. The WS small-world with $p = 0.08$, though having a cost lower than that of the GBA, has a higher average power consumption and a quite the same degree of synchronization for high values of the diffusion coefficient.

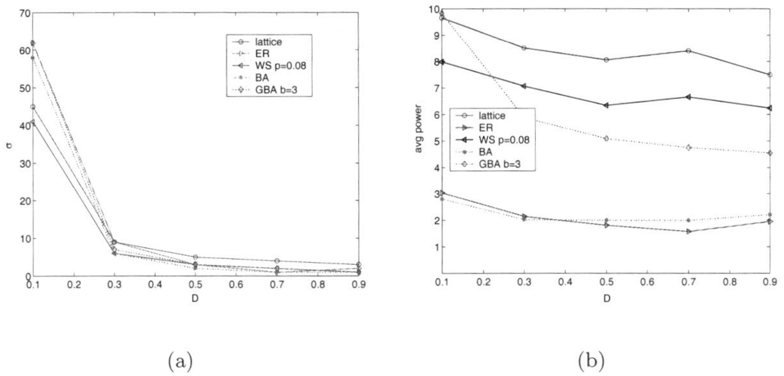

Fig. 2.18 (a) Synchronization index and (b) average power consumption for five different topologies with $N = 196$ nodes, average degree $< k >= 4$, and electrical synapses as functions of the diffusion coefficient D.

We take into consideration an intermediate value of the diffusion coefficient ($D = 0.3$) and investigate the properties of the GBA model as functions of the parameter b (Fig. 2.19). The synchronization index and the average power consumption increase with the exponent b. Nevertheless, this increasing is widely counterbalanced by the remarkable reduction of the cost in the region around $b = 3$, that seems to be a suitable region.

We now examine the case of neuron networks with chemical synapses as in equations (2.21)-(2.25) with a value of $E_{syn} = 1.52$ (excitatory chemical synapses).

The results obtained are quite different from those obtained with the electrical synapse. The lattice is the topology that reaches the lowest values of the synchronization index. The ER random graph, on the contrary, is not able to reach a value smaller than 16. Coherently, the WS model, being similar to the lattice but with some long range connections (typical of the ER), shows values of the synchronization index σ that are slightly greater than those of the lattice.

The difference in the behavior of the two kinds of synapses seems to be found in the different role of the degree. In the electrical synapse, the diffusive effect tends to equalize the potential of connected neurons, independently of the number of neighbors. In the use of chemical synapses the net input a neuron receives from synaptic neurons emitting synchronized spikes is proportional to the number of connected units (and thus to its

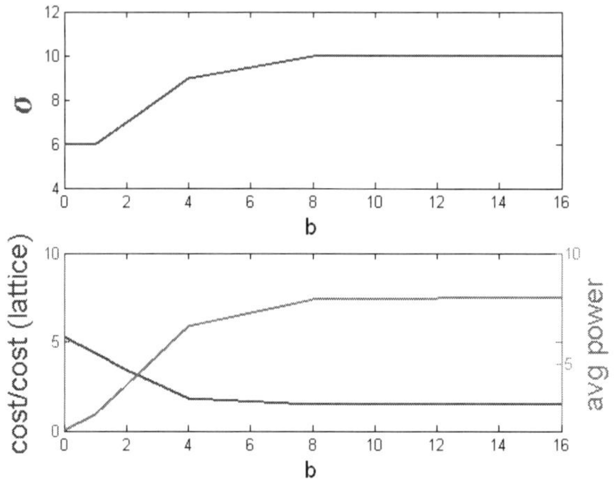

Fig. 2.19 Synchronization index (in the upper part), structural cost and average power consumption (in the lower part) for the Generalized BA model with electrical synapses as function of b.

degree). Hence, for chemical synapses, if all the nodes in a network have the same degree (as in the case of the lattice), synchronization will be enhanced, while, if different nodes have different degree (as in the case of the ER random graph), synchronization will be hampered.

It is not surprising that the BA and the GBA are able to reach values of the synchronization index lower than those of the ER. In fact, though these networks, having a wide range of variation of the degree, are very heterogeneous, it should not be forgotten that there are many nodes that have the same low degree. It is the existence of these nodes that helps synchronization.

In Fig. 2.20 the synchronization index and the average power consumption are plotted for all the analyzed networks. It is worth to note that the two topologies that show better synchronization (lattice and WS) have an optimal value of the maximum conductance $g_{synMax} = 0.8$, for which the synchronization index is minimum. In fact, a further increasing of the conductance not only produces an increase of the average power consumption, but also deteriorates the synchronization degree.

An analogous behavior is shown in Fig. 2.21 for the same models as before, but with an average degree $< k >= 8$. The only difference is that the optimal value of the maximum conductance g_{synMax} for the lattice and

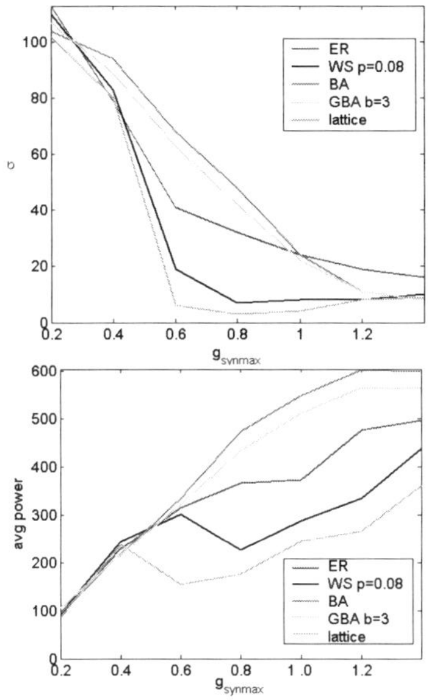

Fig. 2.20 Synchronization index and average power consumption as functions of the maximum conductance g_{synMax} for: Lattice, Erdös-Rényi random graph, Watts and Strogatz small-world model with $p = 0.08$, Barabási-Albert and Generalized Barabási-Albert models, all with chemical synapses.

the WS model is reduced to 0.4. As further confirmation, we compare in Fig. 2.22, the synchronization index and the average power consumption for the lattice with average degrees $< k >= 4$ and $< k >= 8$.

From the analysis discussed we can conclude that in neuron networks synchronization shows to be strictly dependent on the kind of synapse used in the networks. In fact, in the case of electrical synapses, the models that show a better trade-off between high synchronization and low structural and energetic costs are those with both good global and local properties: among the static networks the WS and among the dynamical ones the GBA models. On the contrary, in the case of chemical synapses, there is not a real trade-off, because the lattice, thanks to its regularity, is the network with best synchronization, lowest structural cost and lowest average power consumption.

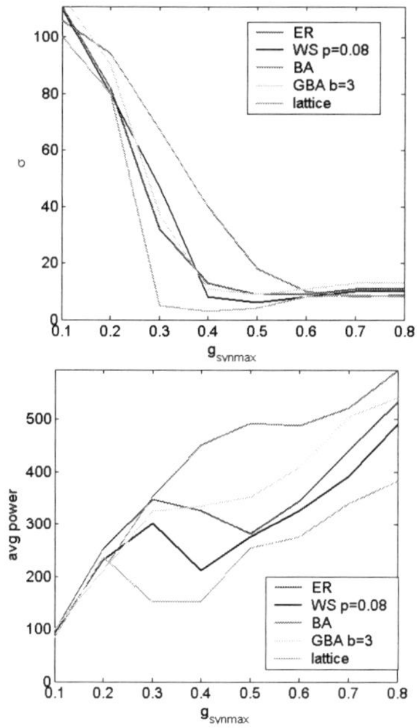

Fig. 2.21 Synchronization index and average power consumption as functions of the maximum conductance g_{synMax} for: Lattice, Erdös-Rényi random graph, Watts and Strogatz small-worls model with $p = 0.08$, Barabási-Albert and Generalized Barabási-Albert models, all with average degree $< k >= 8$ and with chemical synapses.

2.6 Neurons in Presence of Noise

Real neurons constantly operate in presence of noise. To study the effects of noise in networks of neurons, numerical simulations of different topologies of populations of FHN neurons have been carried out. Each neuron is subjected to an independent source of noise. In these conditions the behavior of the population depends on the connections among the elements. By analyzing several kinds of topology (ranging from regular to random) different behaviors have been observed. Several topologies behave in an optimal way with respect to the range of noise level leading to an improvement in the stimulus response coherence, while other ones with respect to the maximum values of the performance index. However, the best results in terms of both the suitable noise level and high stimulus response coherence have been obtained when a diversity in neuron characteristic parameters

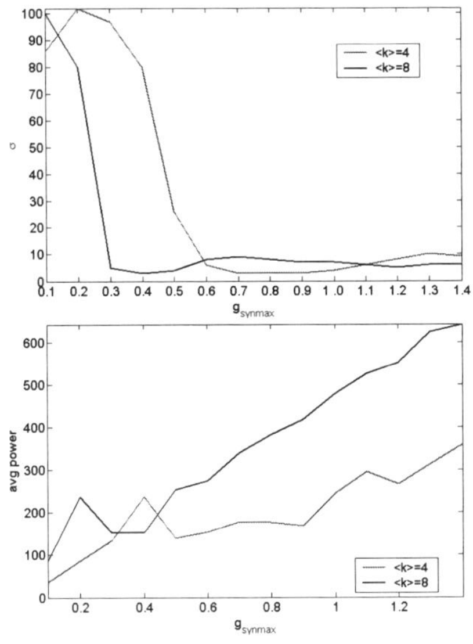

Fig. 2.22 Synchronization index and average power consumption as functions of the maximum conductance g_{synMax} for the lattice with average degrees $< k >= 4$ and $< k >= 8$, and with chemical synapses.

has been introduced and the neurons have been connected in a small-world topology.

Although the promising effects of large scale ensembles of neurons in presence of noise have been investigated, only an idealized connectivity has been taken into account in [Collins (1995)]. We extended the work to several connection topologies and investigate the effects due to the presence of noise in such networks.

The idea that noise usually degrades the performance of a system has been overcome by the concept of stochastic resonance [Benzi (1981); Gammaitoni (1998)]: in nonlinear dynamical systems noise can have benefic effects. Now the concept of stochastic resonance has been extended to account for all the cases in which the presence of noise enhances the degree of order of a system or improve its performance [Anishchenko (2003)]. Examples are the motion of an overdamped Brownian particle in a bistable potential with weak periodic modulation, nonlinear circuits, lasers, biological systems. Precedent studies on stochastic resonance in summing networks of neurons modelled by FitzHugh-Nagumo (FHN) equations have demonstrated that the sensitivity of neuron populations to under-threshold inputs enhances with the number of neurons involved in the computation [Collins (1995)]. The focus of this paper is on the investigation of the role of topology and parameter dissymmetries in neuron networks in presence of noise. We will refer to the enhancement in the stimulus response, due to the presence of noise, as a generalized stochastic resonance effect.

Several network topologies have been investigated focusing on the effects of connections in networks of nonlinear FHN units affected by noise. Structures, like chains or fully connected graphs, random graphs and small-world networks [Strogatz (2003)], have been simulated by connecting FHN neurons excited by weak input signals. The ratios between the input signals and the detected ones have been evaluated to point out the stochastic resonance features of extended neuron populations versus the topology configuration.

The phenomenon studied in this paper is also related to noise-enhanced phase synchronization, that has been introduced in [Zhou (2002)] and applied to spiking systems in [Zhou (2003a,b)]. Noise-enhanced phase synchronization occurs in irregularly spiking systems, where noise can play a constructive role by enhancing phase synchronization.

The equations of the FHN in presence of noise include two inputs, a noise term $\xi(t)$ and a "signal" input $s(t)$, as follows:

$$\begin{cases} \varepsilon \frac{dv_i}{dt} = v_i(v_i - a)(1 - v_i) - w_i + A + s(t) + D\xi_i(t) \\ \frac{dw}{dt} = v_i - w_i - b \end{cases} \tag{2.31}$$

In eq. (2.31) the variable $v_i(t)$ is the fast variable mimicking the membrane potential, $w_i(t)$ is the slow variable modelling the recovery dynamics and A is a constant current bias. Moreover, $s(t)$ represents the input signal and $\xi(t)$ is a Gaussian white random signal. The behavior of the FHN neuron in absence of noise ($D = 0$) can be characterized by a silent state, that in terms of dynamical systems theory is a stable equilibrium point: the system is excitable, when the input signal is chosen over the threshold characteristic of the system a spike is emitted. The FHN neuron shows characteristic signatures of stochastic resonant behavior due to the presence of this threshold-dependent dynamics. It can be clearly identified a noise value optimal in the sense that the system responds to a sub-threshold stimulus with a coherent emission of spikes. This phenomenon allows to detect signals under the threshold and is therefore very important in neurophysiological sensory systems.

Regular network architectures with diffusive coupling have been first investigated. We define r as the radius of the neighborhood, so that by varying r the network architecture changes from local connections to all-to-all global coupling. We also considered the case of uncoupled neurons (represented by $r = 0$). In all the cases the same input has been applied to each element of the population, whereas independent noise source for each neuron has been taken into account. The response to the stimulus of the entire network has been monitored by considering the average of each membrane variable v_i, namely $V(t) = \sum_i v_i(t)$.

Coherence between input and related firing has been quantified by an index taking into account that the information in neural systems is coded in spikes. Therefore, ideally at each input peak should correspond a spike or a spike train. By counting the number of true spikes P_{true} (i.e. spikes in correspondence of an input peak) and false spikes P_{false} (i.e. spikes occurring without any input excitation) the following index is defined:

$$C = \left| c_1 \frac{P_{true}}{P_{tot}} - c_2 \frac{P_{false}}{P_{tot}} \right| \tag{2.32}$$

where P_{tot} is the total number of input peaks. In general the system may respond to a given stimulus peak with a burst of spikes. The weights c_1 and c_2 take into account the density of true spikes elicited by the same input peak and the density of false spikes, respectively. More precisely, let us indicate with N_{burst} the number of spikes in a burst of spikes elicited by the same input peak, and with N_{true} and N_{false} the number of true and false spikes in a burst, respectively. The coefficients c_1 and c_2 are given by $c_1 = \frac{N_{true}}{N_{burst}}$ and $c_2 = \frac{N_{false}}{N_{burst}}$. It should be noticed that $C = 1$ when the spikes and input peaks are coherent or perfectly in anti-phase. The calculation of the index C has been performed by evaluating the coherence between the common input $s(t)$ and $V(t)$.

In the case in which the neurons are not coupled [Collins (1995)], a low sensitivity of large summing networks with respect to the noise level emerges. This phenomenon is referred to as stochastic resonance without tuning. However, the hypothesis of uncoupled neurons is far from the biological case of neuron assemblies, where coupling is the natural solution to enhance information exchange and robustness within the network. For this reason we investigate the emergence of stochastic resonance without tuning in the case of regular networks.

Architectures of globally coupled neurons show different behaviors according to the value of the coupling strength. In Fig. 2.23(a), which refers to low coupling values ($J = 0.01$), performance similar to the case of uncoupled neurons has been observed. By increasing the value of the coupling to $J = 0.03$ as shown in Fig. 2.23(b), the index assumes higher values in spite of a reduction of the range of noise leading to stochastic resonance. This phenomenon is due to the fact that the coupling strength reflects the amount of information exchanged among neurons that leads to collective behavior. As more information is exchanged, the coherence of the response to the stimulus is enhanced, thus leading to higher values of the index. On the other hand, stochastic resonance without tuning is a prerogative of the presence of many sources of noise acting on independent neurons. Thus, the attitude to synchronization, that is higher for increasing values of the coupling, has the effect of enhancing the performance of the system and its sensitivity to the noise. In fact, the minimum noise level detectable is higher in the case of $J = 0.03$.

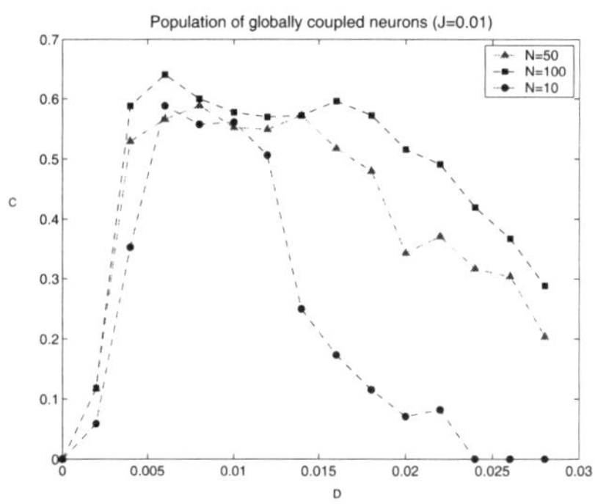

(a)

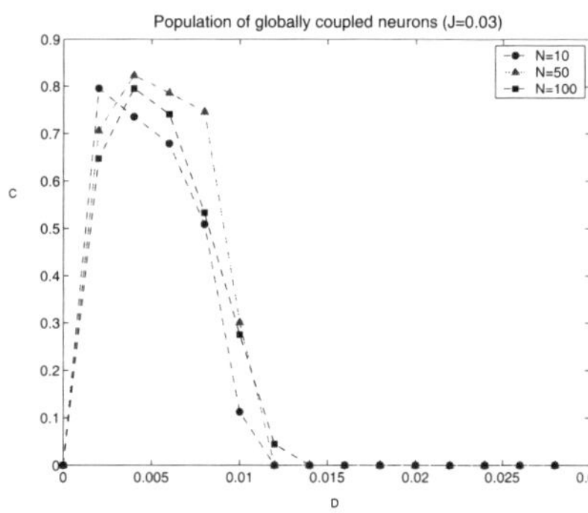

(b)

Fig. 2.23 Values of the index C for a population of globally coupled FHN neurons: (a) $J = 0.01$; (b) $J = 0.03$.

Further simulations have been carried out by keeping constant the value of the coupling strength and investigating the effect of increasing the neighborhood radius starting from the case of nearest neighbor coupling (i.e. $r = 1$). In Fig. 2.24 three cases are reported: $r = 1$, $r = 3$ and $r = 5$. The performance increases according to increasing values of the radius r. It has been verified that starting from a given value of r the performance does not significantly increase.

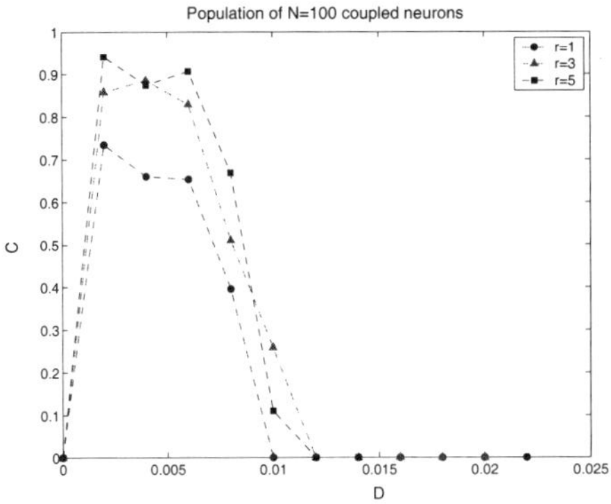

Fig. 2.24 Values of the index C for a population of locally coupled FHN neurons with $r = 1$, $r = 3$ and $r = 5$.

Regular architectures, so far investigated, represent an abstraction of real neuron assemblies, but allow to focus on the dynamical features of the population in a simple connection pattern. On the opposite side, random networks are characterized by asymmetric structures and capture in an idealized way the features of many real systems. These networks have a short average path length which derives from the fact that, starting from each node, any other node of the network can be reached in a small number of links [Strogatz (2003)].

Many real networks (as for instance social networks) have a short average path length, but at the same time show a high clustering degree due to the presence of both short-range and long-range links. In order to model these systems Strogatz and Watts introduced the concept of small-world networks that successfully captures, for instance, the essential features of the neuronal system of the C. Elegans [Strogatz (2003)]. Small-world networks can be built starting from a network of locally coupled neurons and replacing some links with new random ones with probability p. By increasing the probability p the architectures of the neuron population is tuned between the two extremes, regular and random networks. Small-world networks are characterized by low values of the probability p.

We considered small-world networks as random networks and furthermore we remove the hypothesis that all the neurons are identical. In fact, in recent works [Bucolo (2002)] it has been demonstrated that parametric and structural dissymmetries may enhance the synchronization properties of spatially extended networks. For example, it has been shown how an array of slightly different pendula (or Josephson Junctions) organizes itself, while a disordered behavior is observed in an array of identical elements. Moreover, dissymmetries obtained by using deterministic processes have been compared with random ones. These deterministic dissymmetries are obtained starting from a chaotic process and exploiting the noise-like spreadband spectrum of a chaotic signal. A chaotic state variable is sampled to generate a sequence of unpredictable values. Such a sequence can be used to introduce dissymmetry in a system in two main ways: parameter and structural dissymmetries. In a large number of cases the introduction of a deterministic fluctuation in the parameters of the units constituting the complex system has lead to an improvement of the self-organizing properties. On the other side deterministic sequences may also be used to generate the structure of connections in a small-world topology.

Both cases have been investigated in FHN populations. The results are summarized in Fig. 2.25, where are also reported the cases of uncoupled neurons (i.e. the best case as regards the range of suitable noise levels) and random connections (i.e. the best case as regards the maximum level of stimulus response coherence). The introduction of structural dissymmetries has several advantages: with respect to the case of random connections, it increases the range of suitable noise and mantains a high value of the maximum level of C. But the improvement given by the introduction of an unhomogeneity in the parameters of each single neuron is even greater. When non identical FHN neurons arranged in a small-world network (regulated by a deterministic structural dissymmetry) are taken into account, the network behaves optimally both with respect to the maximum level of C and to the range of suitable noise. In particular, the results have been obtained by considering A as the parameter accounting for the unhomogeneity of the neuron population and letting it variable in the range of $[-0.01,\ 0.01]$, while the chaotic sequence has been generated by using the peak-to-peak dynamics (see Chapter 4 or [Strogatz (2000)]) of the variable x of a Double Scroll Chua's Attractor.

From the analysis of networks of noisy neurons, we can conclude that systems behave in a very different way. For example, when uncoupled neurons are considered, the network exhibits stochastic resonance without tuning. The same behavior occurs in network of locally coupled neurons only if the coupling strength is low.

On the other hand, increasing the coupling of regular topologies enhances the performance of the network and decreases the minimum value of noise leading to stochastic resonance. Moreover, the overall range of

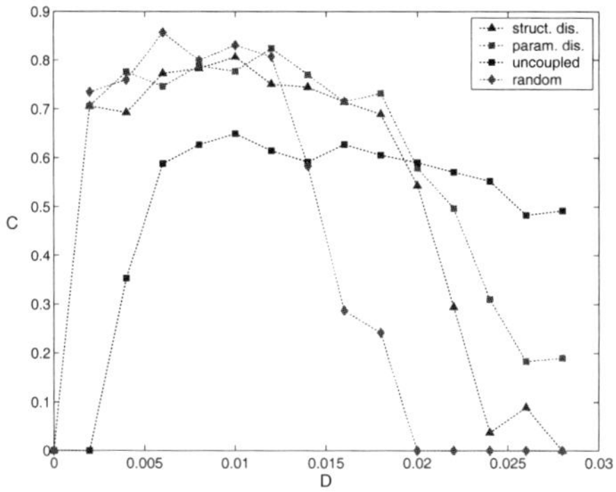

Fig. 2.25 Comparison between different architectures of FHN populations: with structural dissymmetry (▲), with parametric dissymmetry (•), random (♦) and uncoupled (■).

suitable noise level to observe stochastic resonance is reduced. Neuron populations with random connections are those characterized by the highest values of the stimulus response coherence. The simulation results lead to the conclusion that connections improve the performance of the system by increasing the information exchange among the neurons, but need a tuning of the noise level. When non identical FHN neurons arranged in a small-world network (regulated by a deterministic structural dissymmetry) are taken into account, the network behaves optimally with respect to both the aspects of the stimulus response.

2.7 Conclusions

In this chapter, neuron models are briefly reviewed and several complex systems made of elementary units modelling neuron dynamics are then investigated. Several cases are taken into account: the case in which neurons are diffusively coupled into a lattice is first investigated by applying the theory of Local Activity. Then, the case of networks with coupling between neurons based on either electrical or chemical synapses is considered, leading to the conclusion that the behavior of the whole system strongly depends on the topology adopted and on the synapse considered. Finally, the behavior of networks of neurons in presence of noise is discussed. Electronic implementations of both single cell dynamics and groups of neurons are also considered.

References

Abarbanel H. D. I., Rabinovich M. I., Selverston A., Bazhenov M. V., Huerta R., Sushchik M. M. and Rubchinskii L. L., SYNCHRONISATION IN NEURAL NETWORKS. Physics–Uspekhi 39 (4) (1996) 337–362.

Anadigm: www.anadigm.com.

Anishchenko V.S., Astakhov V., Neiman A., Vadivasova T., Schimansky-Geier L., *Nonlinear Dynamics of Chaotic and Stochastic Systems*, Springer Series in Synergetics, 2003.

Arena P., Bucolo M., Fazzino S., Fortuna L., Frasca M., 2005, "The CNN Paradigm: Shapes and Complexity", *Int. J. Bifurcation and Chaos*, Vol. 15, No. 7, pp. 2063-2090.

Arena P., Fortuna L., Frasca M., Giaquinta A., 2000, "A New Circuit for Neuron Implementation", *Proceedings of 2000 International Symposium on Nonlinear Theory and applications (NOLTA2000)*, vol. 1, pp. 397-400.

Barabási A.-L. and Albert R., "Emergence of scaling in random networks", *Science* 286 (1999) 509.

Benzi R., Sutera A., Vulpiani A., "The mechanism of stochastic resonance", *J. Phys. A*, **14** (1981) L453.

Bove M., Grattarola M., Massobrio G., Giugliano M., Martinoia S. "Dynamics of networks of biological neurons: simulation and experimental tool" in *Neural Network System Techniques and Applications, Algorithms and Architectures* - vol. 1, Leondes, CT ed., Academic Press, San Diego, (1998), pp. 401–423.

Bucolo M., Caponetto R., Fortuna L., Frasca M., Rizzo A., "Does chaos works better than noise?", *CAS Magazine* **2**, 3 (2002) 4–19.

Caponetto R., Fazzino S., Fortuna L., Frasca M., 2004, "E^3: An universal emulator for complex systems", *AIP Proceedings of 8th Experimental Chaos Conference 2004, Florence, Italy 14-17 June 2004*, pp. 301-306.

Chua L. O. and Roska T., 2002, *Cellular Neural Networks and Visual Computing: Foundation and Applications*, Cambridge University Press.

Chua L. O., *CNN: A Paradigm for Complexity*, World Scientific, 1998.

Chua L. O., 1989, "Passivity and complexity", *IEEE Trans. Circuits and Systems I*, vol. 46, no. 1, pp. 71-81.

Collins J.J., Chow C. C., Imhoff T. T., "Stochastic resonance without tuning", *Nature*, **376** (1995) 236–238.

Cosenza S., Crucitti P., Fortuna L., Frasca M., La Rosa M., Stagni C., Usai L., "From net topology to synchronization in Hindmarsh-Rose neuron grids", *Mathematical Biosciences and Engineering*, Vol. 2, No. 1, pp. 53-77, 2005.

Dogaru R., 2003, *Universality and Emergent Computation in Cellular Neural Networks*, World Scientific Series on Nonlinear Science, Series A, vol. 43.

Dogaru R. and Chua L. O., 1998, "Edge of Chaos and Local Activity Domain of FitzHugh-Nagumo Equation", *Int. J. Bifurcation and Chaos*, vol. 8, no. 2, pp. 211-257.

Dogaru R. and Chua L. O., 1998, "Edge of Chaos and Local Activity Domain of Brusselator CNN", *Int. J. Bifurcation and Chaos*, vol. 8, no. 6, pp. 1107-1130.

Dogaru R. and Chua L. O., 1998, "Edge of Chaos and Local Activity Domain of Gierer-Meinhardt CNN", *Int. J. Bifurcation and Chaos*, vol. 8, no. 12, pp. 2231-2340.

Erdös P. and Rényi A., "On Random Graphs", *Publicationes Mathematicae*, 6 (1959) 290.

Fortuna L., La Rosa M., Nicolosi D., Sicurella G., "Spatio-temporal dynamics towards self-synchronization index", *Proc. of the XII Int. IEEE Workshop on Nonlinear Dynamics of Electronic Systems*, Évora, Portugal, 2004.

Gammaitoni L., Hnggi P., Jung P., Marchesoni F., "Stochastic resonance", *Mod. Phys. Rew.*, 70, 23, (1998).

Giaquinta A., Argentina M., Velarde M. G., 2000, "A simple generalized excitability model mimicking salient features of neurons dynamics", *J. Stat. Phys.* 101(1/2), 665-678.

Haken H., *Synergetics, An Introduction*, Springer Series in Synergetics, SpringerVerlag, Heidelberg, Berlin, New York, 1983.

Hindmarsh J. L. and Rose R. M., "A model of the nerve impulse using two first order differential equations" *Nature*, 296, pp. 162-164, 1982.

Hindmarsh J. L. and Rose R. M., "A model of neuronal bursting using three coupled first order differential equations", *Proc. R. Soc. Lond. B*, 221, 87, pp. 87-102, 1984.

Hodgkin A. L., "The local electric changes associated with repetitive action in a non-medulated axon", *J. Physiol.* vol. 107, pp. 165-181, 1948.

Izhikevich E. M., "Neural Excitability, Spiking and Bursting", *Int.J.Bif and Chaos*, Vol. 10, no. 6, 2000.

Izhikevich E. M., "Simple Model of Spiking Neurons", *IEEE Transactions on Neural Networks*, Vol. 14, No. 6, 1569–1572, 2003.

Langton C. G., "Computation at the Edge of Chaos: Phase Transition and Emergent Computation", *Physica D*, 42, pp.12-37, 1990.

Min L., Crounse K. R. and Chua L. O., 2000, "Analytical Criteria for Local Activity and Applications to the Oregonator CNN", *Int. J. Bifurcation and Chaos*, vol. 10, no. 1, pp. 25-71.

Nicolis G. and Prigogine I., *Self-Organization in Non-Equilibrium Systems*, Wiley, 1977.

Ottino J. M., "Engineering complex systems", *Nature*, 427, pp. 399, 2004.

Rinzel J., Ermentrout G.B., "Analysis of neural excitability and oscillations", eds. C. Koch, I. Segev, *Methods in Neuronal Modeling*, (The MIT Press, Cambridge), 1989.

Schrodinger E., *What is Life? The Physical Aspect of the Living Cell*, Cambridge University Press, 1948.

Shepherd G. M., *Neurobiology*, Oxford Univ. Press, 1997.

Strogatz S.H., "Exploring complex networks", *Nature* **410** (2001).

Strogatz S.H., *Nonlinear Dynamics and Chaos*, Perseus Publishing, 2000.

Yang T., Chua L. O., 2001, "Testing for Local Activity and Edge of Chaos", *Int. J. Bifurcation and Chaos*, vol. 11, no. 6, pp. 1495-1591.

Zhou C.S., Kurths J., Allaria E., Boccaletti S., Meucci R., Arecchi F.T., "Noise-Enhanced Phase Synchronization of Chaotic Oscillators", *Physical Review Letters*, **89**, 1, (2002), 014101.

Zhou C.S., Kurths J., Allaria E., Boccaletti S., Meucci R., Arecchi F.T., "Noise-enhanced synchronization of homoclinic chaos in a CO2 laser", *Physical Review E* , **67** (2003), 015205 R.

Zhou C.S., Kurths J., Allaria E., Boccaletti S., Meucci R., Arecchi F.T., "Constructive effects of noise in homoclinic chaotic systems", *Physical Review E* , **67**, (2003), 066220.

Chapter 3

Frequency Analysis and Identification in Atomic Force Microscopy

M. Basso and D. Materassi

3.1 Introduction

In the last few years nanotechnologies have had a considerable development due to the realization of devices capable of obtaining images with a nanometric resolution. The Atomic Force Microscope (AFM) is one of the most diffused instruments of this kind [Meyer (2003)]. The concept of force microscopy is based on the measurement of forces between a sharp tip and a sample surface . The tip is mounted on the end of a cantilever which acts as a force sensor. Depending on the operating mode, either the static deflection of the cantilever or the change in its dynamic properties due to the tip-sample interaction can be exploited.

In this chapter, we limit the study to the so-called "tapping" or "dynamic", whose schematic is depicted in Fig. 3.1: the cantilever is periodically forced by a piezo placed under its support inducing a periodic oscillation that is naturally influenced by the interaction forces between the cantilever tip and the sample to analyze. The topography can be inferred by slowly moving the cantilever along the sample surface by means of a piezoactuator and by measuring the amplitude of the cantilever deflection through an optical lever method. A feedback controller driving the piezo input voltage is employed to reject variations of the separation between the sample and the tip center of oscillation due to the sample topography. To this aim, the separation-amplitude curve, that is the plot of the oscillation amplitude as a function of the above separation, is a crucial information in order to retrieve the tip-sample distance from the amplitude signal along the scanning lines. Moreover, topography is not the only information one

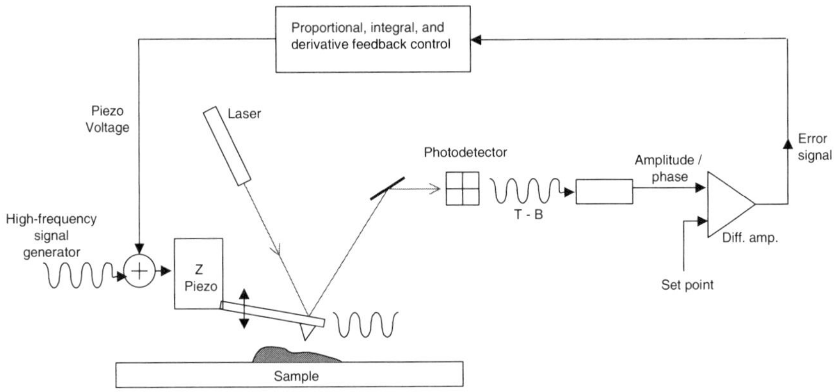

Fig. 3.1 Schematic of a tapping-mode AFM.

might be interested in. In many biological applications, other sample features and characteristics are even more important than imaging, and are strictly related to the identification of the tip-sample interaction forces. Unfortunately, these forces have a highly nonlinear behavior depending on a number of factors, such as different materials properties, different surface forces, operating conditions, etc., which affect the separation-amplitude relation and can even generate complex dynamics (see, e.g., [Ashab (1999); Basso (2000)]). These phenomena make the important problem of identifying different surface forces quite challenging.

In practice, the standard methods to infer some of the sample characteristics from specific experiments are one of the following:

(1) directly retrieving static force-distance curves;
(2) computing force-distance curves from separation-amplitude ones.

While the first approach has many limitations, specially when dealing with soft biological samples, the latter is generally based on the identification of a tip-sample interaction force model for the AFM. Many such models are available in literature (see, e.g., [Cappella (1999)]), although in most cases the separation-amplitude curve is generally obtained by computationally intensive numerical simulations. In this respect, only a few models are capable to provide interesting analytical results. For instance, in [Salapaka (2000)] and [Materassi (2003)] an analytical expression is obtained for two simplified classes of models.

In this work we analytically evaluate the separation-amplitude curve for a large class of models comprising some of most common interaction potential functions studied in the literature, such as the classical Lennard-Jones potential [Cappella (1999)]. The proposed approach relies on the use of harmonic balance techniques [Atherton (1981)], providing an useful insight of many nonlinear known phenomena with respect to fully numerical approaches. Other peculiar features of the proposed class of models are: *i)* can easily account for energy losses by incorporating the hysteresis function approach developed in [Materassi (2003)]; *ii)* is suited for nonlinear frequency-domain identification techniques such as those proposed in [Salapaka (2001)] and [Basso (1997)] .

The paper is organized as follows. In Section 3.2 the considered AFM model class is described; in Section 3.3 the Harmonic Balance (HB) technique is exploited for the analysis of the AFM tip oscillations, whereas in Section 3.4 identification methods and some preliminary experimental results are provided. Finally, in Section 3.5 main results are briefly discussed.

3.2 AFM Modeling

AFM cantilevers operating in dynamic mode are almost always modeled as a feedback interconnection of a SISO linear system $\mathcal{L}$ and a nonlinear static function h as depicted in Fig. 3.2. Models of this kind are well-known in

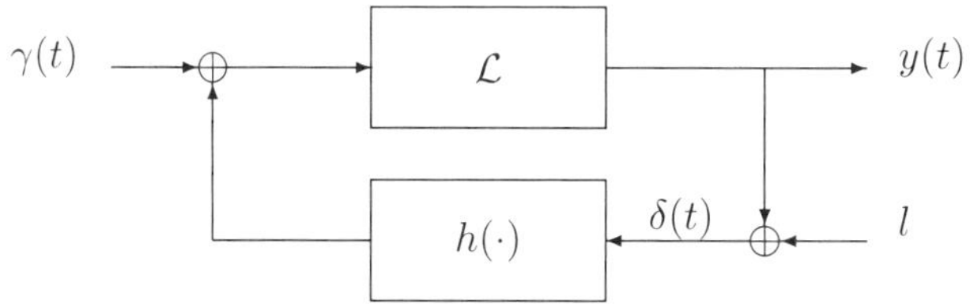

Fig. 3.2 A feedback interconnection of a linear system and a nonlinear static function.

control literature and are usually referred to as Lur'e models [Khalil (1996)]. Using a symbolic notation, we can write the system equation in the following form

$$L\left(\frac{d}{dt}\right) y(t) = h[\delta(t)] + \gamma(t) \tag{3.1}$$

where $L(s)$ is the transfer function of the system $\mathcal{L}$, $y(t)$ is the measured output (that is the cantilever tip position), l is the separation, $\delta(t) = y(t)+l$ represents the tip-sample distance, and $\gamma(t)$ is the external periodic forcing

$$\gamma(t) = \Gamma \cos \omega t \quad .$$

From a physical point of view, the transfer function $L(s)$ describes the free cantilever dynamics only, while the nonlinear function h represents the interaction between the tip and the sample. AFM cantilevers have their lengths in the range of $100 - 300\mu m$. When operating in dynamic mode, common values for the tip oscillation amplitudes are of the order of $10 - 100nm$, that is less than 10^{-3} times the whole cantilever length. Thus, it seems reasonable to assume a linear behaviour for an AFM cantilever, which, practically, acts as simple elastic beam. The identification of $L(s)$ can be achieved indipendently from h in many manners. The most popular techniques perform this identification task measuring the system response to the thermal noise or a simple frequency sweep excitation when the sample is absent [Gibson (2005)].
In Fig. 3.3 the result of an identification based on the thermal response is shown.

Conversely, modeling the tip-sample interaction force h is still a challenging task. The main difficulty lies in the choice of a suitable class of functions to describe the force potential. In fact, microscopic bodies are subject to reciprocal forces that are strongly repulsive at short distances and weakly attractive at long ones. In addition to such a highly nonlinear behaviour, in dynamic mode it has also some relevance to take into account dissipations and energy losses onto the sample surface.
Force curves are traditionally obtained by moving the sample relative to the cantilever in a quasi-steady state manner when the cantilever is not forced. The sample is moved slow enough that it can be assumed the cantilever tip settles to the equilibrium point corresponding to the given separation l and previous initial conditions. Then, the measured data can be fitted using a proper class of functions. However, such "static" force curves do not characterize the tip-sample interaction behavior when the cantilever is oscillating. In fact, they cannot describe the damping and lossy processes of the sample, as those dissipation phenomena typically depend on the cantilever velocity and are absent when the velocity is zero. Instead, it

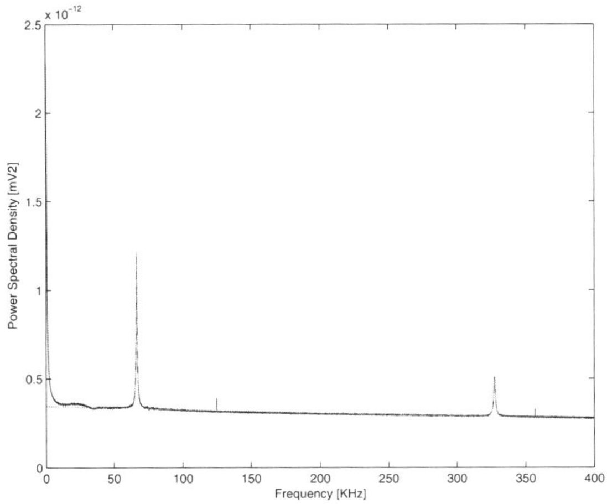

Fig. 3.3 The system thermal response (solid line) is very well fitted by a fourth order model (dashed line).

is reasonable to think that the interaction force could show a distinct behaviour when then tip is continuosly moving, as it happens in the dynamic mode. This motivates the development of tools to perform a "dynamic" identification in order to model the forces acting on the cantilever, using functions depending on the tip velocity $\dot{y} = \dot{\delta}$, too.

We consider the following class of functions

$$h(\delta, \dot{\delta}) = \begin{cases} \sum_{n=1}^{N} K_n^- h_n(\delta) & \text{if } \dot{\delta} < 0 \\ \\ \sum_{n=1}^{N} K_n^+ h_n(\delta) & \text{if } \dot{\delta} \geq 0 \end{cases} \tag{3.2}$$

where $h_n(\delta)$ are a base of suitable non-negative functions and the dependence on $\dot{\delta}$ occurs in h considering only its sign. From a physical perspective, we can consider (3.2) as a simple hysteresys law made of two different positional forces: the first one acts when the tip and the sample are approaching and the second one when they are getting further. In order to actually make the interaction described by $h(\delta, \dot{\delta})$ dissipative, we should also impose some additional constraints on the parameters K_n^+ and K_n^-. For example, the condition $K_n^- \geq K_n^+$ is sufficient since it makes every base element dissipative by itself (although it is not a necessary one).

In our analysis, we will consider two special cases of this hysteretic interaction since they allow to obtain analytically or, at least, computationally efficient results.

3.2.1 *Piecewise interaction force*

The first class of potential functions we treat contains the functions $h(\cdot)$ in the form (3.2) where $N = 2$ and

$$h_n(\delta) = \begin{cases} 0 & \text{if } \delta \geq 0 \\ \delta^{n-1} & \text{if } \delta < 0 \end{cases} \tag{3.3}$$

In Fig. 3.4 the shape of such a kind of functions is depicted. The variable δ in this model represents the distance between the tip and the region where the sample begins the interaction.

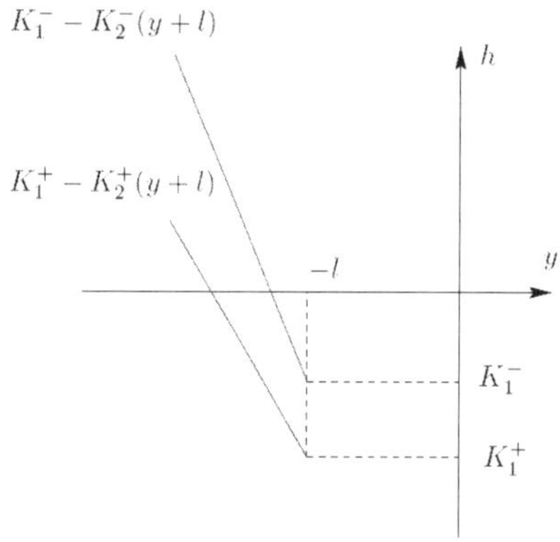

Fig. 3.4 Interaction force modeled by a piecewise linear function.

3.2.2 *Lennard Jones-like interaction force*

The adoption of the Lennard-Jones potential

$$h(\delta) = \frac{K_n}{\delta^n} + \frac{K_m}{\delta^m} \qquad n, m \in \mathcal{N}; K_n, K_m \in \mathcal{R} \tag{3.4}$$

is a quite common choice when fitting statically measured curves [Cappella (1999)]. Such potential indicates long range attractive forces and short range strong repulsive forces acting on the tip when $n > m$ and $K_m < 0 < K_n$. As analogy, we consider a possible generalization of the Lennard-Jones potential in the form (3.2) where

$$h_n(y) = \frac{1}{\delta^n}. \tag{3.5}$$

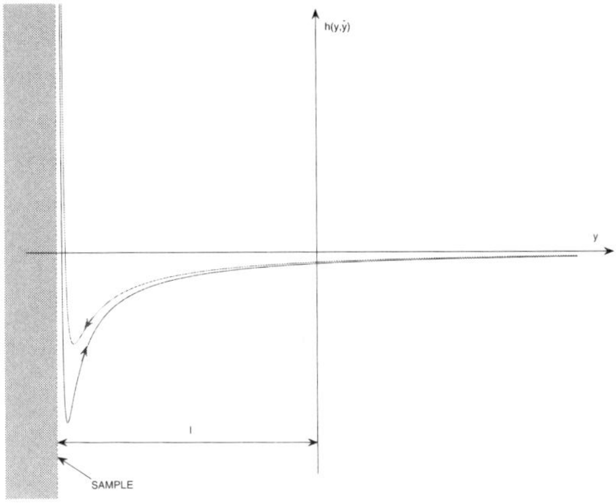

Fig. 3.5 Sketch of a Lennard-Jones-like interaction with hysteresis.

The variable δ now represents the tip distance from the sample region where the interaction force becomes infinitely large (see Fig. 3.5).

As it is obvious, apart an additive constant the two "δ"s in the proposed classes of models have identical meaning and roles.

3.3 Frequency Analysis via Harmonic Balance

Considering physical values for the mass and the elastic constant of the cantilever, the linear part of the Lur'e system given by $L(s)$ in (3.1) typically shows a marked filtering effect beyond the first resonance peak due to a very small damping. This suggests to exploit the harmonic balance (HB) method to analyze the system periodic behavior [Atherton (1981); Khalil (1996)]. Limiting to the first order harmonic, we can assume $y(t) \simeq \mathrm{Re}[A + B\mathrm{e}^{i\omega t}]$ and approximate the corresponding output of the nonlinear hysteresis block $h\big(y(t) + l, \dot{y}(t)\big) \simeq \mathrm{Re}[N_0 A + N_1 B\mathrm{e}^{i\omega t}]$, where

$$\begin{cases} N_0 = N_0(A, B, \omega) := \dfrac{1}{A}\dfrac{1}{T}\displaystyle\int_0^T h\big(y(t), \dot{y}(t)\big)\mathrm{d}t \\ N_1 = N_1(A, B, \omega) := \dfrac{1}{B}\dfrac{2}{T}\displaystyle\int_0^T h\big(y(t), \dot{y}(t)\big)\mathrm{e}^{-i\omega t}\mathrm{d}t \end{cases} \tag{3.6}$$

are the constant and harmonic gains of the nonlinear block usually denoted as the describing functions of the nonlinearity. For the general class of hysteretic force models introduced, we obtain

$$N_0 = \frac{1}{2\pi A}\sum_{n=1}^{N}\left(\int_{-\pi}^{0} K_n^+ h_n(l+A+B\cos\tau)\mathrm{d}\tau + \int_{0}^{+\pi} K_n^- h_n(l+A+B\cos\tau)\mathrm{d}\tau\right)$$
$$= \frac{1}{2\pi A}\sum_{n=1}^{N}\Sigma_n \int_{0}^{+\pi} h_n[B(q+\cos\tau)]\mathrm{d}\tau$$

where

$$\Sigma_n := K_n^- + K_n^+$$

and

$$q := \frac{l+A}{B}\ . \tag{3.7}$$

Similarly, we find for N_1

$$N_1 = \frac{1}{\pi B}\sum_{n=1}^{N}\left(\int_{-\pi}^{0} K_n^+ h_n[B(q+\cos\tau)]\mathrm{e}^{-i\tau}\mathrm{d}\tau + \int_{0}^{+\pi} K_n^- h_n[B(q+\cos\tau)]\mathrm{e}^{-i\tau}\mathrm{d}\tau\right)$$
$$= \frac{1}{\pi B}\sum_{n=1}^{N}\left(\Sigma_n \int_{0}^{+\pi} h_n[B(q+\cos\tau)]\cos\tau\mathrm{d}\tau - i\Delta_n \int_{0}^{+\pi} h_n[B(q+\cos\tau)]\sin\tau\mathrm{d}\tau\right) \tag{3.8}$$

where

$$\Delta_n := K_n^+ - K_n^-\ .$$

The periodic solutions of system (3.1) with a sinusoidal forcing $\gamma(t) = \mathrm{Re}[\Gamma\mathrm{e}^{i(\omega t+\phi)}]$ can be computed through the classical describing function method which originates the following equation to be solved in A, B, ϕ

$$A + B\mathrm{e}^{i\omega t} = -L(0)N_0 A + L(i\omega)[-N_1 B + \Gamma\mathrm{e}^{i\phi}]\mathrm{e}^{i\omega t} \qquad \forall t \tag{3.9}$$

or, equivalently,

$$\begin{cases} [1 + L(0)N_0(A,B)]A = 0 \\ [1 + L(i\omega t)N_1(A,B)]\,B = L(i\omega t)\Gamma\mathrm{e}^{i\phi}. \end{cases} \tag{3.10}$$

Finally, we can easily decouple the variable ϕ from (3.10)

$$\begin{cases} [1 + L(0)N_0(A,B)]A = 0 \\ |1 + L(i\omega t)N_1(A,B)|\,B = |L(i\omega t)|\Gamma. \\ \phi = \arg\left[L(i\omega)^{-1} + N_1(A,B)\right] \end{cases} \tag{3.11}$$

3.3.1 *Piecewise interaction model analysis*

For the piecewise-linear potential described in Section 3.2.1 we obtain

$$\begin{cases} N_0 = \frac{1}{A}\left[\Sigma_1 R_1(q) + \Sigma_2 R_2(q) B\right] \\ N_1 = \frac{1}{B}\left[\Sigma_1 S_1(q) + i\Delta_1 T_1(q)\right] + \left[\Sigma_2 S_2(q) + i\Delta_2 T_2(q)\right] \end{cases} \tag{3.12}$$

where

$$R_1(q) := \frac{\operatorname{acos}(q)}{2\pi} \qquad R_2(q) := \frac{q\operatorname{acos}(q) - \sqrt{1-q^2}}{2\pi}$$

$$S_1(q) := -\frac{\sqrt{1-q^2}}{\pi} \qquad S_2(q) := \frac{\operatorname{acos}(q) - q\sqrt{1-q^2}}{2\pi}$$

$$T_1(q) := \frac{1-q}{\pi} \qquad T_2(q) := -\frac{(1-q)^2}{2\pi}$$

Finally, by the substitutions

$$\chi(q) := \Sigma_1 R_1(q) \qquad \Omega(q) := \Sigma_2 R_2(q)$$

$$\Phi(q) := \Sigma_1 S_1(q) + i\Delta_1 T_1(q) \qquad \Psi(q) := \Sigma_2 S_2(q) + i\Delta_2 T_2(q)$$

we can obtain a formally handy expression for the describing functions

$$N_0 = \frac{1}{A}\left[\chi(q) + \Omega(q) B\right]$$

$$N_1 = \frac{\Phi(q)}{B} + \Psi(q).$$

In this model, the variable q can represent the "penetration" of the tip in the sample. In fact, assuming as exact the first harmonic approximation, we have that for $q > 1$ the tip does not get in contact with the sample; for $q = 1$ the tip grazes the sample and for $q < 1$ we have an effective interaction. The case for $q < -1$ is meaningless in this model since it would indicate that the tip completely oscillates "into" the sample; so we will not need to analyze it deeply.

From (3.10), it is also possible to formally write B as a function of q only. In fact

$$B = \frac{\Gamma}{|L(i\omega)^{-1} + N_1|} \tag{3.13}$$

implies

$$\left|L(i\omega)^{-1}B + BN_1\right|^2 = |L(i\omega)^{-1}B + \Phi + \Psi B|^2 = \Gamma^2. \tag{3.14}$$

The substitutions $\hat{\Phi} := \Phi$ and $\hat{\Psi} := \Psi + L(i\omega)^{-1}$ yield

$$\begin{aligned} &(\hat{\Phi} + \hat{\Psi}B)(\hat{\Phi}^* + \hat{\Psi}^* B) = \Gamma^2 \Rightarrow \\ &|\hat{\Psi}|^2 B^2 + 2\mathrm{Re}[\hat{\Phi}\hat{\Psi}^*]B + |\hat{\Phi}|^2 - \Gamma^2 = 0 \end{aligned} \tag{3.15}$$

which is a simple second order algebraic equation whose roots are

$$B = \Theta(q) := \frac{-\mathrm{Re}[\hat{\Phi}\hat{\Psi}^*] \pm \sqrt{\mathrm{Re}[\hat{\Phi}\hat{\Psi}^*]^2 - |\hat{\Psi}|^2(|\hat{\Phi}|^2 - \Gamma^2)}}{|\hat{\Psi}|^2}. \tag{3.16}$$

Substituting in (3.11) and reminding that $l = qB - A$, we can finally write

$$\begin{cases} A(q) = -L(0)\left[\chi(q) + \Psi(q)\Theta(q)\right] \\ B(q) = \Theta(q) \\ \phi(q) = \arg\left[L(i\omega)^{-1} + N_1(A, \Theta(q))\right] \\ l(q) = q\Theta(q) + L(0)\left[\chi(q) + \Psi(q)\Theta(q)\right] \end{cases} \tag{3.17}$$

The variable q, for its definition (3.7), depends on A, B and l, therefore equations (3.17) are only implicit relations, "formally masked" as explicit ones. System (3.17) can not be solved in closed form since it involves transcendental equations. However, it is possible to obtain its solution through a conceptually easy method. Since the variable l is a known parameter of the model, it is possible by the last of (3.17), to determine the corresponding values of q and then A, B and ϕ by exploiting the remaining equations. In other words, we have transformed the problem of solving the whole system (3.10) into the easier problem of solving a single real equation in the unknown q.

Experimentally, the separation-amplitude curve is obtained slowly moving the sample towards the cantilever and measuring both the amplitude of the first harmonic and the separation. Although it is not possible to derive an explicit analytical form for $B = B(l)$, we can give a parametric form for it. By using the "q-explicit" equations in (3.17) we can consider the parametric curve

$$\begin{cases} l = l(q) \\ B = B(q) \end{cases} \qquad \forall\, q \in \mathbf{R}. \tag{3.18}$$

In Fig. 3.6 we can observe the comparison beetween simulated results and the predicted ones using the analytical expression (3.18). The two curves show a good agreement. In addition, we note that there is a range of values of l for which three different periodic solutions exist, two stable and one unstable. This critical behavior, exhibited for some values of the parameters, generates "jumps" depending on the scan direction (l increasing or decreasing). Both experiments and simulations confirm this phenomenon [Kyhle (1997)]. The proposed approach has allowed, in addition, to detect the unstable periodic solutions which are difficult to obtain by simulation. A similar behavior can be observed in the phase-separation

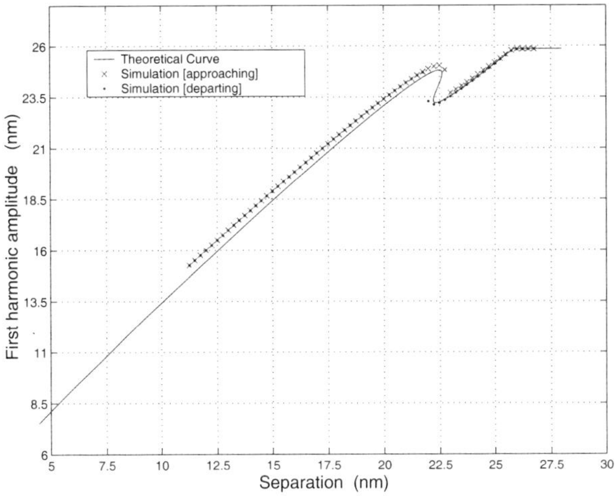

Fig. 3.6 Comparison beetween the separation amplitude curve obtained analytically and by a simulation.

diagram and can be explained exactly in the same way. In addition, even in this simple model, we can observe that a relation of "quasi-linearity" exists between amplitude and separation [Salapaka (2000)].

3.3.2 *Lennard Jones-like hysteretic model analysis*

For the generic hysteretic interaction force of the class (3.5), we can again evaluate the describing functions N_0 and N_1 of the nonlinearity h:

$$\begin{cases} N_0 = \sum_{n=1}^{N} \frac{\Sigma_n}{AB^n} R_n(q) \\ N_1 = \sum_{n=1}^{N} \frac{1}{B^{n+1}} \left[\Sigma_n S_n(q) + i\Delta_n T_n(q)\right] \end{cases}$$

where the functions

$$\begin{aligned} R_n(q) &:= \frac{1}{2\pi}\int_0^\pi \frac{1}{(q+\cos\tau)^n}\mathrm{d}\tau \\ S_n(q) &:= \frac{1}{\pi}\int_0^\pi \frac{\cos\tau}{(q+\cos\tau)^n}\mathrm{d}\tau \\ T_n(q) &:= \frac{1}{\pi}\int_0^\pi \frac{-\sin\tau}{(q+\cos\tau)^n}\mathrm{d}\tau. \end{aligned}$$

can be analitically evaluated for any given n and $q > 1$.

Imposing harmonic balance we get

$$\begin{cases} A = -L(0)\sum_{n=1}^{N}\Sigma_n\frac{R_n(q)}{B^n} \\ \left[\Gamma e^{i\phi} - \sum_{n=1}^{N}\frac{\Sigma_n S_n(q)+i\Delta_n T_n(q)}{B^n}\right] L(i\omega) = B. \end{cases} \tag{3.19}$$

The second equation of (3.19) can be expressed in the form

$$L(i\omega)\Gamma e^{i\phi} = L(i\omega)\sum_{n=1}^{N}\frac{\Sigma_n S_n(q) + i\Delta_n T_n(q)}{B^n} + B \tag{3.20}$$

We can remove ϕ by multiplying each term by its conjugate. Finally, multiplying by B^{2N} the equation can be easily rewritten as a $(N+2)$-degree polynomial in the variable B whose coefficients depend on the variable q only

$$p(B) = \sum_{n=1}^{2N+2} C_n(q)B^n = 0. \tag{3.21}$$

It can be easily shown that $C_{2N+2} = 1/|L(i\omega)|^2$, $C_{2N+1} = 0$ and $C_{2N} = -\Gamma^2$. For sufficiently large q (that is when the interaction is neglegible) we have that $C_k \cong 0$, $\forall k < 2N$, such that

$$p(B) \cong (|L(i\omega)|^{-2}B^2 - \Gamma^2)B^{2N} = 0 \tag{3.22}$$

One evident root is $B \cong \Gamma|L(i\omega)|$ with the meaning that, if the cantilever tip is far away from the sample, a periodic solution is detected and its first harmonic amplitude is about $l_0 = \Gamma|L(i\omega)|$, the free oscillation amplitude.

For every $q > 1$, the polynomial equation (3.21) can be efficiently solved in B keeping only solutions with a "physical meaning" (the real and positive ones). Then, the constant component of the periodic solution A can be easily evaluated exploiting the first of (3.19). The phase ϕ can also be similarly obtained as a function of the parameter q

$$\phi(q) = \arg\left\{L^{-1}(i\omega) + \sum_{n=1}^{N}\frac{\Sigma_n S_n(q) + i\Delta_n T_n(q)}{B^{n+1}}\right\}. \tag{3.23}$$

Finally, the parameter l is given by the original relation

$$l(q) = qB(q) - A(q). \tag{3.24}$$

The final result is that the variables A, B, ϕ and l are all expressed with respect to the parameter q. The separation-amplitude diagram can be therefore obtained considering the pair $(l(q), B(q))$ which describes a curve in a parametric form. A similar procedure can be used to obtain the relation between any two variables.

The proposed analysis technique is computationally efficient for this class of models such that it easily allows to investigate the influence of model parameters in the system behavior.

3.4 Identification of the Tip-Sample Force Model

In this section we develop a method to identify the parameters defining the proposed interaction models directly from experimental data.

3.4.1 *Identification method*

As previously discussed, the transfer function $L(s)$ can be indipendently estimated, so it assumed known. A, B, ϕ are quantities which can be immediately evaluated by the measured signal $y(t)$. Moreover l is a known quantity (apart an additive offset which can be easily estimated). Thus, the variable $q = \frac{A+l}{B}$ can be assumed known, too.
The class of functions (3.2) chosen to model the interaction has the useful property to be linear in its parameters K_n^- and K_n^+. This allows to develop a simple identification procedure which makes again use of HB techniques. Let us define The first order harmonic balance equation leads to a set of $2N$ linear equations in the unknown variables Σ_m and Δ_n

$$\begin{cases} \Gamma\cos(\phi) - \mathrm{Im}[L^{-1}(i\omega)]B = \sum_{n=1}^{N} \Sigma_n \frac{S_n(A,B)}{B^{n+1}} \\ \Gamma\sin(\phi) - \mathrm{Re}[L^{-1}(i\omega)]B = \sum_{n=1}^{N} \Delta_n \frac{T_n(A,B)}{B^{n+1}} \end{cases} \tag{3.25}$$

Assuming to have $M > 2N$ experimental measurements $(A_1, B_1, \phi_1, l_1)...(A_M, B_M, \phi_M, l_M)$ and adopting a more compact notation, we can write two indipendent matrix equations

$$\begin{aligned} P_S\Sigma &= Q_S \\ P_D\Delta &= Q_D \end{aligned} \tag{3.26}$$

where

$$\Sigma := \begin{pmatrix} \Sigma_1 \\ \vdots \\ \Sigma_N \end{pmatrix} \qquad \Delta := \begin{pmatrix} \Delta_1 \\ \vdots \\ \Delta_N \end{pmatrix} \tag{3.27}$$

are the unknown vectors and

$$\begin{aligned} P_S[m,n] &:= \frac{S_n(A_m,B_m)}{B_m^{n+1}} \quad & Q_S[m] &:= \Gamma\cos(\phi) - \mathrm{Im}[L^{-1}(i\omega)]B_m \\ P_D[m,n] &:= \frac{T_n(A_m,B_m)}{B_m^{n+1}} \quad & Q_D[m] &:= \Gamma\sin(\phi) - \mathrm{Im}[L^{-1}(i\omega)]B_m \end{aligned} \tag{3.28}$$

are constant matrices.
Σ and Δ can be evaluated from (3.26) using a least-square method. The presence of linear constraints on the unknown parameters (for example to force the identified force to be dissipative) can be easily handled, too. In fact, the problem just reduces to a contrained quadratic optimization.

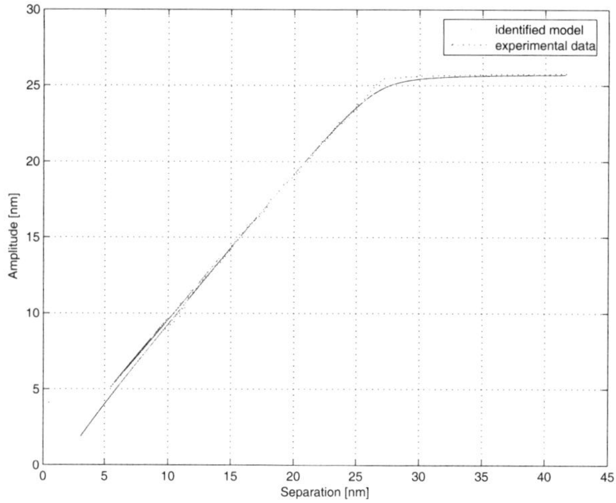

(a)

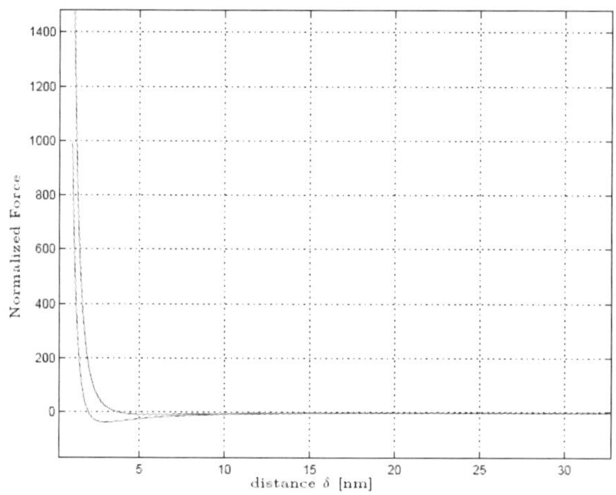

(b)

Fig. 3.7 An experimental separation-amplitude curve (a) fitted using a Lennard-Jones like model ($n_a = 2$, $n_r = 3$) for the interaction force (b).

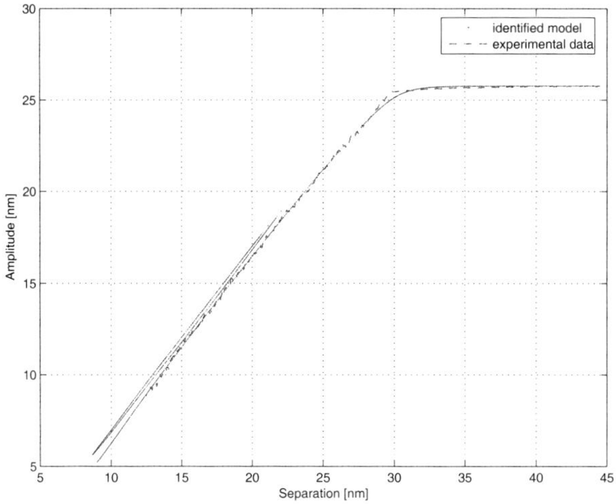

(a)

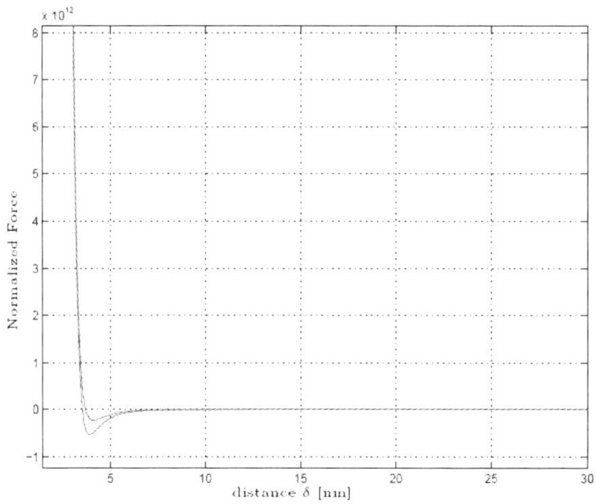

(b)

Fig. 3.8 An experimental separation-amplitude curve (a) fitted using a Lennard-Jones like model ($n_a = 6$, $n_r = 13$) for the interaction force (b).

3.4.2 Experimental results

A simplified interaction model compared to (3.2) and (3.5) has been considered

$$h(y,\dot{y}) = \begin{cases} \frac{K^-_{n_a}}{\delta^{n_a}} + \frac{K^-_{n_r}}{\delta^{n_r}} & \text{if } \dot{\delta} < 0 \\ \\ \frac{K^+_{n_a}}{\delta^{n_a}} + \frac{K^+_{n_r}}{\delta^{n_r}} & \text{if } \dot{\delta} \geq 0 \end{cases} \qquad \text{with } n_r > n_a \tag{3.29}$$

The choice of this simplified model is motivated by simplicity and by an analogy with the standard Lennard-Jones potential, where only two main terms are present, accounting for repulsive and attractive forces when $K_{n_a} < 0 < K_{n_r}$. The identification technique described in the previous section has been employed on a set of experimental data using different values for n_a and n_r. The transfer function $L(s)$ has been previously identified using a thermal-response based approach. The results obtained for the two cases $n_a = 2, n_r = 3$ and $n_a = 6, n_r = 13$ are reported in Fig.s 3.7 and 3.8, respectively.

It can be noted that the quality of the identification improves increasing the values of the two esponents n_a and n_r.

3.5 Conclusions

In this chapter we have proposed a class of models for tip-sample interacion dynamics in atomic force microscopy via impact dynamics. The use of a hysteresis law can be well combined with harmonic balance techniques for the analysis of oscillatory behaviors, providing interesting analytical results. For instance, the presence of jump phenomena discovered in many experiments is well-predicted and explained. The suggested method is based on a first order harmonic approximation and gives good quantitative results since the linear part of the considered Lur'e system has a strong filtering effect. In such a situation, the HB technique has advantages over standard numerical approaches since it requires a computational effort much smaller than the one required by a pure simulation of the same model. Moreover, another advantage stems from the ability to identify the model of the tip-sample interaction forces directly from experimental data available in the frequency domain.

References

Ashab M., Salapaka M., Dahleh M., Mezic I., "Dynamical analysis and control of micro-cantilevers", *Automatica*, 35, pp. 1663-1670, 1999.

Atherton D.P., *Stability of Nonlinear Systems*, Wiley, New York, 1981.

Basso M., Giarré L., Dahleh M., Mezic I., "Complex dynamics in a harmonically excited Lennard-Jones oscillator: Microcantilever-sample interaction in scanning probe microscopes," *ASME Journal of Dynamics Systems, Measurement and Control*, 122, pp. 240-245, 2000.

Basso M., Genesio R., Tesi A., Torrini G., "On describing systems with periodic behaviour in terms of simple nonlinear models", *Proc. of the IEEE Conference - Control of Oscillations and Chaos*, St. Petersburg, 1997.

Cappella B., Dietler G., "Force distance by atomic force microscopy', *Surface Science Reports*, 34, pp. 1-104, 1999.

Gibson C., Smith D., Roberts C., "Calibration of silicon atomic force microscope cantilevers", *Nanotechnology*, 16, pp. 234-238, 2005.

Khalil H., *Nonlinear Systems*, Second Edition, Prentice-Hall, Upper Saddle River, 1996.

Kyhle A., Sorensen A., J. Bohr, "Role of attractive forces in tapping tip force microscopy", *Journal of Applied Physics*, 81, pp. 6562-6569, 1997.

Materassi D., Basso M., Genesio R., "A Model for Impact Dynamics and its Application to Frequency Analysis of Tapping-Mode Atomic Force Microscopes", Proc. of IEEE Int. Conference on Decision and Control, Maui, 2003.

Meyer E., Hug H., Bennewitz R., "Scanning Probe Microscopy: The Lab on a Tip", Springer, Berlin, 2003.

Salapaka M., Chen D., Cleveland J., "Linearity of amplitude and phase in tapping-mode atomic force microscopy", *Physical Review B*, 61, pp. 1106-1115, 2000.

Sebastian A., Salapaka M., Chen D., Cleveland J., "Harmonic and power balance tools for tapping-mode AFM", *Journal of Applied Physics*, Vol. 89, pp. 6473-6480, 2001.

Chapter 4

Control and Parameter Estimation of Systems with Low-Dimensional Chaos – The Role of Peak-to-Peak Dynamics

C. Piccardi and S. Rinaldi

A method for deriving a reduced model of a continuous-time dynamical system with low-dimensional chaos is discussed. The method relies on *Peak-to-Peak Dynamics*, namely on the possibility of approximately (but accurately) predicting the next peak amplitude of an output variable from the knowledge of at most the two previous peaks. The reduced model is a simple one-dimensional map or, in the most complex case, a set of one-dimensional maps. The paper illustrates two fields of application of the reduced model: control design and parameter estimation. Attention is paid to the impact of the noise and of the sampling frequency (i.e. the rate of data acquisition) on the performance of the control system and on the quality of the parameter estimation.

4.1 Introduction

Model reduction is certainly one of the most thoroughly studied topics in applied mathematical modeling, as witnessed by hundreds of papers published in the past decades, mostly in the area of linear systems [Obinata (2001)]. In this paper, we discuss a model reduction technique for continuous-time systems that is strictly specific to chaotic systems . The key point is the existence of *Peak-to-Peak Dynamics* (PPD) [Candaten (2000)], i.e. the possibility of approximately (but accurately) predicting the next peak of a scalar output variable (i.e. the amplitude of the next relative maximum) from the knowledge of the last peak value or, in the most complex cases, of the last two peaks. If PPD exist, the system can be described, regardless of its dimension, by a reduced model that consists of one (or a few) one-dimensional maps. The reduced model can be fruitfully used for designing a controller, or for estimating the parameters of an "a priori" assigned ODE description of the system.

The existence of PPD is crucially related to the low-dimensionality of the chaotic attractor [Strogatz (1994); Candaten (2000)]. Although many examples have been pointed out in all fields of science and engineering (see e.g. [Lorenz (1963)] and [Abarbanel (1997)] for an early and a recent contribution), PPD have generally been considered as a curious hallmark of chaos or, at most, as a descriptive tool. Only in recent years the study of PPD has been proposed as a systematic tool for deriving reduced models, which can then be used for control and parameter estimation purposes.

This chapter is a review of several recent contributions in the field [Candaten (2000); Piccardi (2000, 2003)]. In Sec. 2 the conditions giving rise to PPD are discussed in detail, and the structure of the reduced model is analyzed. In Sec. 3, a control problem is formulated and solved taking advantage of the reduced model: the influence of the noise and of the sampling frequency on the controller performance is discussed in detail. The use of the reduced model for the purpose of parameter estimation is illustrated in Sec. 4. The concluding remarks (Sec. 5) briefly touch upon some other extensions.

4.2 Peak-to-Peak Dynamics

We begin by considering an autonomous, continuous-time, finite-dimensional system of the form

$$\dot{x}(t) = f(x(t)), \tag{4.1}$$

where $t \in R_+ = \{t \geq 0\}$ is time, $x \in R^n$ is the state vector, and $f : R^n \to R^n$ is a smooth function. Assume that (4.1) has a single attractor $X \subset R^n$ which is a *chaotic attractor*, i.e. an invariant, attractive set exhibiting sensitive dependence on initial conditions (e.g. [Guckenheimer (1983); Eckmann (1985); Alligood (1996)]). Thus $x(t)$ is aperiodic but bounded. The system is observed through a scalar output variable

$$y(t) = g(x(t)), \tag{4.2}$$

where $g : R^n \to R$ is a smooth function.

Now suppose that $x(0) \in X$ (i.e. neglect the transient toward the attractor) and denote by t_k and y_k, respectively, the time instant and the amplitude of the k-th relative maximum (*peak*) of $y(t)$ $(0 < t_1 < t_2 < \ldots)$, i.e. $y_k = y(t_k)$, $\dot{y}(t_k) = 0$, $\ddot{y}(t_k) < 0$ (provided the maximum is quadratic). Then define the *Peak-to-Peak Plot* (PPP) as the set $S = \{(y_k, y_{k+1}), k = 1, 2, \ldots\}$, i.e. the set of all ordered pairs of consecutive peaks. Figure 4.1 shows four PPP's obtained through the simulation of four different chaotic dynamical systems: Lorenz system [Lorenz (1963)], a chemical reactor [Peng (1990)], Chua circuit [Pivka (1996)], and Rössler hyperchaotic system [Rössler (1979)]. Although the sets S displayed in Fig. 4.1 are all fractal sets, i.e. they have non-integer dimension, in the examples (a), (b) and (c) they can be accurately approximated by a suitable set of smooth curves. By contrast, this cannot be done in example (d). In the first three cases, we say that *system (4.1)-(4.2) has PPD.*

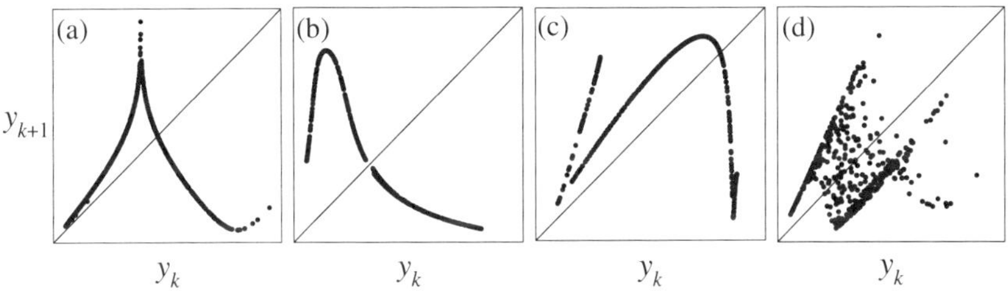

Fig. 4.1 Four Peak-to-Peak Plots. (a) Lorenz system. (b) Chemical reactor. (c) Chua circuit. (d) Rössler hyperchaotic system.

Once a fitting criterion has been specified, the PPP of a system with PPD defines a (possibly multiple-valued) function

$$y_{k+1} = F(y_k). \tag{4.3}$$

The PPD are said to be *complex* when F is actually a multiple-valued function (i.e. multiple values of y_{k+1} are associated to some y_k), and *simple* otherwise. Thus Fig. 4.1 shows two examples of simple PPD ((a) and (b)) and one of complex PPD ((c)). In this work, for simplicity, we consider only systems with simple PPD.

It is rather easy to understand that the condition under which system (4.1)-(4.2) has PPD is $\dim(X) \cong 2$, where $\dim(X)$ is the fractal dimension (e.g. capacity dimension) of the chaotic attractor X. Indeed, denote by

$x_k = x(t_k)$ the state corresponding to the k-th output peak, i.e. $y_k = g(x_k)$. Note that $x_k \in \Sigma$, where Σ is the Poincaré section defined by $\dot{y} = 0$, i.e. the manifold

$$\sum_{i=1}^{n} \frac{\partial g(x)}{\partial x_i} f_i(x) = 0. \tag{4.4}$$

Introducing the set $P = \{x_k, k = 1, 2, \ldots\}$ (see Fig. 4.2), it turns out that if $\dim(X) \cong 2$, then $\dim(P) = \dim(X) - 1 \cong 1$, so that P can be effectively approximated by a curve segment (or a few curve segments) $P^* \subset \Sigma$. This amounts to replace P, which has non-integer dimension, with the one-dimensional set P^* or, equivalently, to assume that $x_k \in P^*$.

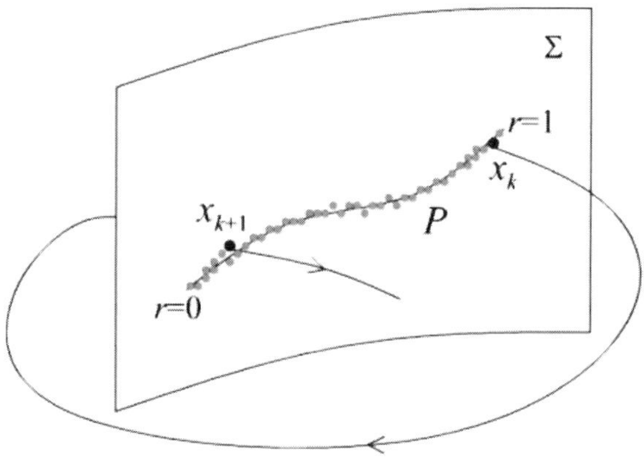

Fig. 4.2 The Poincaré section Σ, with some points of the set P and a segment of the trajectory $x(t)$. The output $y(t)$ has a peak when $x(t) = x_k$ and when $x(t) = x_{k+1}$.

Without loss of generality, a coordinate $r : P^* \to [0, 1]$ can be introduced on P^*. Denoting by $r_k = r(x_k)$ the value associated to x_k, a map $\varphi : [0, 1] \to [0, 1]$ can be defined such that $r_{k+1} = \varphi(r_k)$ since, given x_k, the state x_{k+1} is identified by the Poincaré map $\prod_\Sigma : \Sigma \to \Sigma$, i.e. $x_{k+1} = \prod_\Sigma(x_k)$. On the other hand, denoting by $\rho : [0, 1] \to R$ the restriction of g to P^*, i.e. $y_k = \rho(r_k) = g(x_k)$, we have

$$y_{k+1} = \rho(r_{k+1}) = \rho(\varphi(r_k)). \tag{4.5}$$

Now, if ρ is invertible then $y_{k+1} = \rho(\varphi(\rho^{-1}(y_k)))$, which is an equation of the form

$$y_{k+1} = F(y_k), \tag{4.6}$$

with F single-valued. This is the case of *simple* PPD (see Fig. 4.1(a)-(b), where the knowledge of y_k is sufficient to accurately predict y_{k+1}). The one-dimensional map (4.6) is actually a reduced model of system (4.1)-(4.2) which, regardless of its order n, captures the behavior of the system on its chaotic attractor. From a practical point of view, the map $F : \Omega \rightarrow \Omega$, where $\Omega = [\inf_k y_k, \sup_k y_k]$ is the domain of the peak values y_k, can easily be identified from the PPP by standard fitting techniques.

If, on the contrary, the map ρ is not invertible, more than one value of y_{k+1} is associated to y_k, i.e. F is multiple-valued (see e.g. Fig. 4.1(c)). This is the case of *complex* PPD that, for simplicity, will not be considered in this paper (see [Candaten (2000); Piccardi (2002)] for details).

4.3 Control System Design

The reduced model derived via peak-to-peak analysis can be used to effectively design a controller for the chaotic system . In the previous section, however, the analysis has been restricted to an ideal environment characterized by the absence of noise and by the possibility of sampling the output at arbitrarily high frequency. It is clear that such conditions do not hold in practice, so that it is important to assess whether the model based on PPD is applicable in realistic situations, where the chaotic system is affected by process and measurement noise, and the output is sampled at finite frequency. In other words, instead of continuously measuring the noise-free output $y(t)$, we must realistically assume that a sampled, noise-affected output is available .

The study of the effect of sampling frequency and noise intensity on controlled dynamical systems is a classical topic that has been investigated in recent years in the context of chaotic systems (e.g. [Aguirre (2000); Mees (2000)] and references therein). Such an effect can be very heavy in our case, since the identification of the PPD crucially relies on the ability of correctly detecting and measuring the output peaks. But a too small sampling frequency and/or heavy noise affecting the output can partially or completely hide the "true" peaks of $y(t)$ or even give rise to "false" peaks. Therefore, the possibility of identifying the PPD could be seriously compromised.

The impact of the above issues on the controller design are investigated in this section, with the help of a model of a chemical reactor [Peng (1990, 1991)]. The PPD method for chaos control, in the ideal case, will be illustrated firstly (details can be found in [Piccardi (2000, 2002)]). Then the system will be embedded in a realistic environment, where the output is sampled at finite frequency, and both the state and the output are affected by noise. It will be shown that this can make the peak-to-peak analysis totally inapplicable. Finally, simple filtering and peak-detection techniques will be employed to recover a reliable model of the PPD from noisy data. It will be shown that, provided the noise amplitude and the sampling period are not too large, such a model can be used for designing an effective controller. This ultimately proves that chaos control via PPD has a fair degree of robustness and can therefore be applied to real situations.

4.3.1 *PPD modeling and control*

In this section, we consider continuous-time systems with a (scalar) *control input*, i.e. described by

$$\dot{x}(t) = f(x(t), u(t)), \tag{4.7a}$$

$$y(t) = g(x(t)), \tag{4.7b}$$

where $u \in R$ is the control input. For illustrative purposes, reference will be made to the following third-order model of a chemical reaction [Peng (1990, 1991)]:

$$\dot{x}_1 = u(\alpha + x_3) - x_1 - x_1 x_2^2, \tag{4.8a}$$

$$\dot{x}_2 = \beta(x_1 + x_1 x_2^2 - x_2), \tag{4.8b}$$

$$\dot{x}_3 = \delta(x_2 - x_3). \tag{4.8c}$$

The state variables are (dimensionless) concentrations of reactants, and the control u is the inflow concentration. The parameters are set to $\alpha = 65$, $\beta = 200$, $\delta = 47$, and $y = x_2$ is the output variable.

If the control input is kept constant at a nominal value $u = u_{nom} = 0.1544$, the system displays a chaotic behavior with PPD (Fig. 4.3). In this case, the PPP can be fairly well modeled by a single-value map (*simple* PPD), i.e.

$$y_{k+1} = F(y_k, u_{nom}). \tag{4.9}$$

By considering a few constant values of u in a (small) neighborhood of u_{nom} ($u \in U = [0.1536, 0.1552]$ in the example), a set of PPP's can be derived, so that a set of models (4.9), parameterized in u, can be identified:

$$y_{k+1} = F(y_k, u). \tag{4.10}$$

Each one of these models describes the PPD of system (4.7) in the chaotic regime corresponding to $u(t) = u$. Equation (4.10) can also be used when the control $u(t)$ is piecewise-constant, namely when it is kept constant between two subsequent peaks of $y(t)$, i.e. $u(t) = u_k$ for $t \in (t_k, t_{k+1}]$ (where t_k is the time of occurrence of the peak y_k). Indeed, by replacing u with u_k in (4.10) we obtain

$$y_{k+1} = F(y_k, u_k), \tag{4.11}$$

which is a simplified (i.e. approximate) model of system (4.7) with piecewise-constant control. Some limitations in the use of (4.11) are discussed in [Piccardi (2002)].

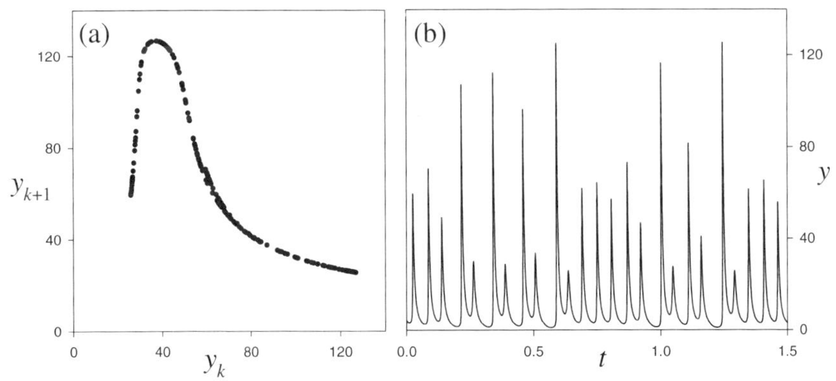

Fig. 4.3 (a) Peak to Peak Plot of system (4.8) with $u = 0.1544$. (b) A segment of the output time series $y(t)$.

If we set a control value $\bar{u} \in U$, a fixed point $\bar{y} = F(\bar{y}, \bar{u})$ of (4.11) corresponds to an *Unstable Periodic Orbit* (UPO) of system (4.7) characterized by one single output peak per period at $\bar{y}$. To stabilize the UPO, the following optimal control problem can be formulated: find the control law $u_k = q(y_k)$, with $u_k \in U$, that minimizes the cost functional

$$J(y_0) = \lim_{N\to\infty} \sum_{k=0}^{N-1} (y_k - \bar{y})^2. \tag{4.12}$$

Such a problem belongs to a class that is extensively treated in the literature [Bertsekas (1995)]. The solution can be found numerically through Dynamic Programming (see e.g. [Piccardi (2000)] for details). Notice that the control law is not constrained to belong to any pre-specified class of functions. In order to control system (4.7) with the control law $u_k = q(y_k)$, the amplitude y_k and the time of occurrence t_k of each peak must be available. Then $u_k = q(y_k)$ is applied to system (4.7) during the interval $(t_k, t_{k+1}]$.

In our example, problem (4.12) has been solved to stabilize the unstable fixed-point $\bar{y} = 62.21$ of (4.11) at $\bar{u} = 0.1544$ ($= u_{nom}$, the central value of the set U). By applying the resulting control law, the system performs as shown in Fig. 4.4. After the controller is switched on at $t = 0.9$, the desired periodic behavior is rapidly reached. Therefore, in the ideal case the control based on the PPD works perfectly.

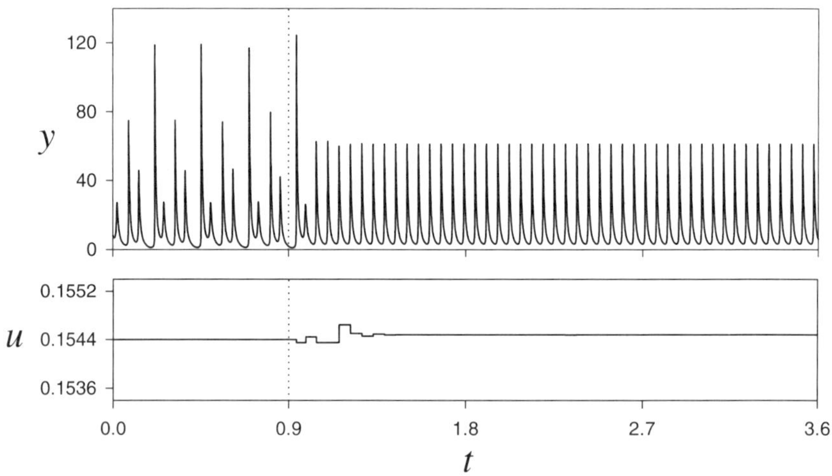

Fig. 4.4 The output y and the input u of system (4.8) controlled by the optimal solution of problem (4.12) (the controller is switched on at $t = 0.9$).

4.3.2 *The impact of noise and sampling frequency*

A set of experiments has been carried out to assess the robustness of the controller with respect to noise intensity and sampling frequency. For that, the model of the chemical reaction (4.8) has been used to generate synthetic output time series, from which the PPP's have been derived.

Three assumptions characterize the experiments:

i) The output $y(t)$ is sampled at constant frequency $1/\tau$ ($\tau > 0$).

ii) The state equation is affected by a multiplicative process noise, i.e.

$$\dot{x}(t) = w(t)f(x(t), u(t)), \tag{4.13}$$

where $w(t)$ is constant during each sampling period, and $(\log w)$ is normally distributed with zero mean and standard deviation σ_w.

iii) The output is affected by multiplicative measurement noise, i.e. the following noise-affected output $z(t)$ is available

$$z(t) = v(t)g(x(t)), \tag{4.14}$$

where $v(t)$ is constant during each sampling period, and $(\log v)$ is normally distributed with zero mean and standard deviation σ_v.

In view of i)-iii), a sampled, noise-affected output $z_h = z(h\tau)$ ($h = 0, 1, \ldots$) is available for model identification and control design. A peak of the sampled time series is detected at time $t = h\tau$ if $z_h > z_{h-1}$ and $z_h > z_{h+1}$. In this way, a sequence $\{z_k\}$ of peaks is associated to the output time series. For simplicity, in all the experiments we have chosen $\sigma_w = \sigma_v (= \sigma)$. A wide set of simulations has been carried out, for several values of the pair (τ, σ) (notice that the ideal situation is recovered for $(\tau, \sigma) \to (0, 0)$). If we denote by τ_{avg} the average time between two subsequent peaks in the noise-free, continuously measured output of system (4.8) ($\tau_{avg} = 0.06$ in our example), three values of the sampling period τ have been considered:

- *High Sampling Frequency*: $\tau_{hf} = \tau_{avg}/100$,
- *Medium Sampling Frequency*: $\tau_{mf} = \tau_{avg}/50$,
- *Low Sampling Frequency*: $\tau_{lf} = \tau_{avg}/10$.

Three values of σ have also been considered:

- *No Noise*: $\sigma = 0$,
- *Low-Level Noise*: $\sigma = \sigma_l = 0.04$ (4% average multiplicative process and output noise),

- *High-Level Noise*: $\sigma = \sigma_h = 0.08$ (8% average multiplicative process and output noise).

It must be stressed that assumptions i)-iii) might prevent the identification of the PPD since, roughly speaking, noise can generate "false" peaks, whereas sampling can make the "true" peaks not detectable with good accuracy. To prove these facts, it is sufficient to consider the PPP's resulting from two particular experiments and compare them with Fig. 4.3. The first one (Fig. 4.5), obtained with $(\tau, \sigma) = (\tau_{lf}, 0)$, shows that, with a low sampling frequency, the peak detection is so inaccurate that PPD cannot be identified even if noise is absent. Indeed, in most situations the peak value detected by the sampled output is considerably lower than the true one. The second PPP (Fig. 4.6) is obtained with $(\tau, \sigma) = (\tau_{hf}, \sigma_l)$, and shows that noise creates a large number of false peaks that obscure the PPD even if the sampling frequency is high. In the next section we will see that simple, heuristic filtering and peak-detection techniques can overcome these drawbacks to a certain extent.

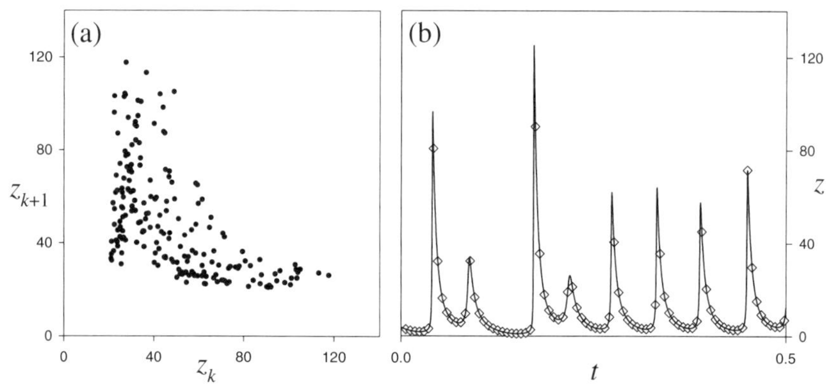

Fig. 4.5 (a) Peak-to-Peak Plot of system (4.13)-(4.14) for $(\tau, \sigma) = (\tau_{lf}, 0)$ (low sampling frequency, no noise), $u = 0.1544$. (b) A segment of the output time series $z(t)$ (solid line) and of the sampled output time series z_h (diamonds).

4.3.3 *PPD reconstruction*

Several heuristic techniques can be used to reconstruct the PPD of a system starting from the sampled output data corrupted by noise. In our case, the

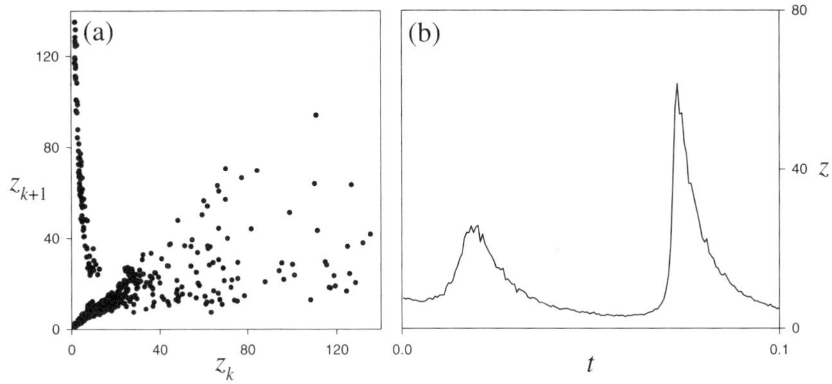

Fig. 4.6 (a) Peak-to-Peak Plot of system (4.13)-(4.14) for $(\tau, \sigma) = (\tau_{hf}, \sigma_l)$ (high sampling frequency, low-level noise), $u = 0.1544$. (b) A segment of the sampled output time series z_h.

best results have been obtained by coupling a moving-average filter with a peak-detection rule based on a moving window. More precisely, let us define the "filtered output" w_h as the (causal) N-point moving average of z_h, i.e.

$$w_h = \frac{z_h + z_{h-1} + \ldots + z_{h-N+1}}{N}. \tag{4.15}$$

Then, a peak w_k of the time series w_h is detected when

$$w_h > \max\{w_{h+i} | i = \pm 1, \pm 2, \ldots, \pm M\}, \tag{4.16}$$

namely when w_h is the greatest value in a $\pm M$ -point test window centered at $t = h\tau$. In this way, a sequence $\{w_k\}$ of peaks is associated to each sampled output record and the PPP is the set of all pairs (w_k, w_{k+1}).

Figure 4.7 presents the set of PPP's obtained through the above described procedure, for the nine pairs (τ, σ) defined in the previous section. For each (τ, σ), the best values of N and M have been found by trial-and-error. The simple inspection of Fig. 4.7 and its comparison with Fig. 4.3 point out that PPD can be reconstructed fairly well with high sampling frequency, even in the presence of high-level noise. The reconstruction is quite good also with medium sampling frequency, even if the PPP with high-level noise leaves a quite considerable uncertainty in predicting w_{k+1}

given w_k. Finally, the reconstruction is practically impossible with low-frequency sampling, even in the case of no noise. In view of this result, low-frequency sampling will no longer be considered in the following.

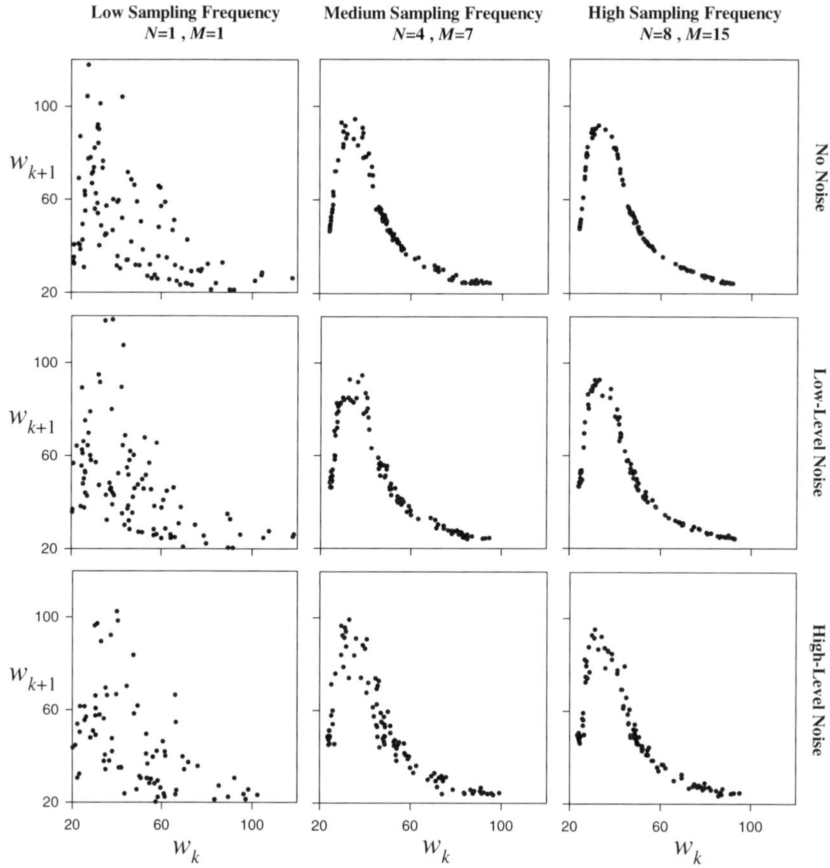

Fig. 4.7 Reconstructed Peak-to-Peak Plots of system (4.13)-(4.14) for nine values of (τ, σ), $u = 0.1544$.

For each sampling frequency and noise level, the same procedure has been used to derive PPP's for the values of $u \in U$ on a uniform grid with $\Delta u = 0.0001$. The result is a set of PPP's $\{(w_k, w_{k+1})\}$, one for each value of u, which has been processed through a standard 3D least-squares interpolation program (TableCurve3D, by Jandel Scientific). This has allowed us to derive an equation of the form

$$w_{k+1} = F(w_k, u), \tag{4.17}$$

which is a model of the form (4.10) governing the peaks of the filtered output w_k. An optimal control problem can be defined with the goal of stabilizing a fixed point of (4.17), which corresponds to a noisy periodic solution of system (4.13)-(4.14). The cost functional (to be minimized) is formally similar to (4.12), namely

$$J(w_0) = \lim_{N\to\infty} \sum_{k=0}^{N-1} (w_k - \bar{w})^2, \tag{4.18}$$

where $\bar{w}$ is a fixed-point of (4.17) for a given input $\bar{u}$, i.e. $\bar{w} = F(\bar{w}, \bar{u})$. The unknown is the control law $u_k = q(w_k)$, i.e. a rule that, once a peak w_k of the filtered output has been detected, prescribes the control u_k to be applied until the next peak w_{k+1}.

It must be pointed out that, if the reconstruction scheme (4.15)-(4.16) is used in real-time, a peak can only be detected with a delay of M sampling intervals (in our examples, 14-15% of the average time interval between two subsequent peaks). Thus, the control is switched to the new value with a considerable delay: in our experiments, this has yielded a poor controller performance. On the other hand, a strong reduction of the test window of equation (4.16) (e.g., to $M/3$) also has given poor results, since the number of false peaks has increased considerably. On the contrary, satisfactory results have been obtained with the following adaptive strategy: while controlling the system, a value w_h is a candidate peak if

$$w_h > \max\{w_{h+i} | i = -M, \ldots, -2, -1, 1, 2, \ldots, P, P < M\}, \tag{4.19}$$

i.e. an asymmetric test window is set up. If (4.19) holds, the control is switched to $q(w_h)$ at time $t = (h + P)\tau$, i.e. with a delay of $P(< M)$ sampling intervals. Subsequently, if $w_{h+1} > w_h$ for some $i = P + 1, P + 2, \ldots, M$, the candidate peak w_h is discarded, and the previous control is restored. Thus, a wrong control is applied for at most $(M - P)$ sampling intervals. In our case, the best results (see below) have been obtained with $P = 1$ for medium sampling frequency and for $P = 5$ for high sampling frequency.

Figure 4.8 presents the results of the simulation of system (4.13)-(4.14) with the optimal control law resulting from problem (4.18), for medium and high sampling frequency , and for low- and high-level noise (as already pointed out, low sampling frequency has not been considered, since the corresponding PPD reconstruction is too poor). In all cases, the control law is able to steer the system toward the prescribed (noisy) periodic solution. This means that the proposed methodology performs satisfactorily. However, in one of the cases (medium sampling frequency and high-level noise), the control system succeeds in steering the system to the prescribed behavior, but only after some time. Moreover, the prescribed behavior is occasionally lost for some time (not shown in the figure).

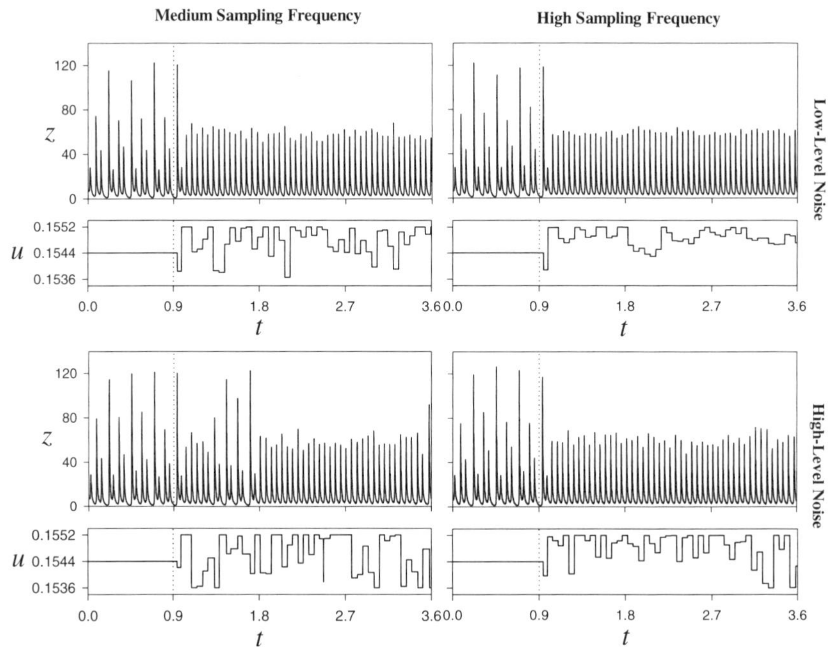

Fig. 4.8 The output z and the input u of system (4.13)-(4.14) controlled by the optimal solution of problem (4.18) (the controller is switched on at $t = 0.9$) for four values of (τ, σ).

In conclusion, it clearly emerges from the experiments that a too low sampling frequency has a more "destructive" effect than high noise levels. Indeed, the PPD of a system affected by noise can be effectively reconstructed by means of very intuitive filtering and peak-detection schemes, provided data are collected at a sufficiently high frequency. But the loss of information due to a too low sampling frequency cannot be recovered. From the control point of view, the analysis has shown that a controller based on PPD has a fair degree of robustness. Indeed, when equipped with a suitable filtering and peak-detection scheme, such a controller performs satisfactorily if the sampling frequency is sufficiently high. It would be interesting to know to what extent other model-free methods for control design of chaotic systems (e.g. those based on embedding techniques) enjoy the same robustness property.

4.4 Parameter Estimation

One of the most important problems in applied mathematical modelling is *identification*, namely the derivation of a model from laboratory or field data. In many situations, the functions defining the model are known, because they are drawn from "first principles" (e.g. the laws of physics) or from empirical assumptions (e.g. logistic growth in ecology, Coub-Douglas production functions in economics). However, some of the parameters entering such functions are unknown and must be estimated from data.

To be more precise, the identification problem has often the following features:

i) The model is assumed to belong to a parameterized set of continuous-time nonlinear systems of the form (1)-(2), i.e.

$$\dot{x}(t) = f(x(t), p), \tag{4.20a}$$

$$y(t) = g(x(t), p), \tag{4.20b}$$

where $p \in P \subset R^m$ is an unknown vector of parameters. The state function $f : R^n \times R^m \to R^n$ and the output function $g : R^n \times R^m \to R$ are assumed to be known.

ii) The output $y(t)$ is evaluated at discrete time instants $0, \tau, \ldots, h\tau, \ldots, (T-1)\tau$, where τ is a fixed *sampling interval.*

iii) The measure at time $t = h\tau$ is affected by noise. In this section we only consider additive measurement noise, i.e. what is actually available is a time series $z_0, z_1, \ldots, z_{T-1}$ defined by

$$z_h = y_h + v_h = y(h\tau) + v_h, \tag{4.21}$$

where v_h is the measurement noise.

Given an output time series $z_0, z_1, \ldots, z_{T-1}$ and the state and output functions f and g, the problem of *parameter estimation* is that of determining the "best" model in the parameterized set (4.20) (i.e. the "best" parameter value $\bar{p} \in P$), according to a specified criterion.

The various strategies for attacking the above problem have been recently surveyed by [Voss (2004)], with particular attention to two classes of methods: *multiple-shooting* algorithms (e.g. [Baake (1992)]), which are mainly designed to overcome the problems of local minima in the optimization cost functional, and recursive parameter estimation techniques *Recursive parameter estimation*, based on Kalman filter and its extensions (e.g. [Julier (2004)]). In this paper, a parameter estimation method for chaotic systems is proposed which is explicitly based on the existence of PPD. The method is rather naïve, as it essentially consists of matching the PPP generated with the parameter-dependent system (4.20) with the PPP derived from the available data. The estimate $\hat{p}$ is the parameter value p which gives rise to the best matching.

The application of the method is rather simple and has several advantages. A reasonable amount of measurement noise is acceptable, provided the data are suitably filtered to correctly construct the PPP. Moreover, the method is insensitive to the state dimension n, provided the system has PPD. On the other hand, the solution is obtained by numerically searching for the minimum of an error function in the m-dimensional parameter space. This function has typically multiple minima, so that finding the optimal value might be problematic, especially if the number of parameters (m) is rather high. However, we will show that fairly good results can be obtained by adopting a suitable minimization strategy, namely by using a genetic algorithm [Goldberg (1989); Gen (2000)] (to identify a promising region in the parameter space) followed by a gradient-like searching algorithm (for local refinement). In the following of this section, the parameter estimation algorithm is illustrated in detail.

4.4.1 *Derivation of the "empirical PPP"*

In Sec. 3 it has been shown that, given the output time series $z_0, z_1, \ldots, z_{T-1}$, the effect of sampling and noise can prevent the correct identification of the PPP since, roughly speaking, noise can generate "false peaks" whereas sampling can make the "true peaks" not detectable with accuracy. Analogously to what has been done for control system design, we use a simple but reliable technique to reconstruct the PPP from the noisy time series, i.e. we define a (non-causal) *filtered output* w_h as the $(2N+1)$-point moving average of z_h, i.e.

$$w_h = \frac{1}{2N+1} \sum_{i=-N}^{N} z_{h+i}, \tag{4.22}$$

and we consider w_h as a peak when it is the greatest value in a $(2M+1)$-point test window centered at time h, i.e. when

$$w_h > max\{w_{h+i} | i = \pm 1, \pm 2, \ldots, \pm M\}. \tag{4.23}$$

In this way, a sequence $\{w_k\}$ of peaks is associated to an output time series, so that we define the "*Empirical PPP*" as the set $S = \{(w_k, w_{k+1}) | k = 1, 2, \ldots, K\}$, where $(K+1)$ is the number of detected peaks.

4.4.2 *Interpolation of the "empirical PPP"*

Since the system, by assumption, has (simple) PPD, the "Empirical PPP" can accurately be interpolated by a curve of the form

$$w_{k+1} = F(w_k), \tag{4.24}$$

where the single-valued function $F : W \rightarrow W$ (with $W = [\inf_k w_k, \sup_k w_k]$) is obtained as the best fitting function (in the least-squares sense) among a given class. In this work, polynomials of a given order q have been used. Of course, the fitting accuracy will be high only in "ideal" conditions (i.e. no noise and very small sampling interval).

4.4.3 *Optimization*

For any given parameter value p, a time series $\hat{y}_0, \hat{y}_1, \ldots, \hat{y}_{T-1}$ is generated by numerically solving (4.20) in $0 \leq t \leq \tau(T-1)$, and by sampling the output at frequency $1/\tau$. Then, a filtered time series is obtained by a (2N+1)-point moving average, i.e.

$$\hat{w}_h = \frac{1}{2N+1} \sum_{i=-N}^{N} \hat{y}_{h+i}, \tag{4.25}$$

and a peak of the series $\hat{w}_h$ is attained when

$$\hat{w}_h > \max\{\hat{w}_{h+i} | i = \pm 1, \pm 2, \ldots, \pm M\}. \tag{4.26}$$

The values of N, M are the same used in (4.22)-(4.23). This procedure defines a "Model PPP", i.e.

$$\hat{S} = \{(\hat{w}_k, \hat{w}_{k+1}) | k = 1, 2, \ldots, \hat{K}\}, \tag{4.27}$$

which has to be interpolated by a q-order polynomial, defining the function

$$\hat{w}_{k+1} = \hat{F}_p(\hat{w}_k). \tag{4.28}$$

Now, the parameter estimate $\hat{p}$ is the value of p that gives rise to a "Model PPP" which is the "closest one" to the "Empirical PPP". In other words, $\hat{p}$ is the solution to

$$\min_p J(p) = \min_p \sum_{k=1}^{K} \left(F(w_k) - \hat{F}_p(w_k) \right)^2 . \tag{4.29}$$

Problem (4.29) must be solved by means of numerical iterative procedures. It is easy to check that $J : R^m \to R$ is typically characterized by multiple local minima. As already said, fairly good results have been by using a genetic algorithm to identify a promising region in the parameter space, followed by a gradient-like searching algorithm for local refinement.

4.4.4 *Example of application*

The method has been tested on several chaotic systems displaying PPD: here we illustrate the application to the chemical reaction model considered in Sec. 3 (the equations are repeated below for convenience). A few data series have been synthetically generated by numerically integrating (4.20) with a nominal parameter value $p = \bar{p}$, and then by corrupting the output samples y_h with a white, gaussian noise v_h with zero mean and standard deviation σ_v given by

$$\sigma_v = \sigma_y \frac{NSR}{100}, \tag{4.30}$$

where σ_y is the standard deviation of y_h, and NSR [%] is a prescribed *noise-to-signal ratio*. The values adopted for NSR have been 0 (no noise), 5 and 10.

For each data series, the parameter estimation procedure described above has been carried out. The quality of the result is evaluated by the (percentage) *total estimation error*

$$e_{tot} = 100\frac{\|\hat{p} - \bar{p}\|_2}{\|\bar{p}\|_2}. \tag{4.31}$$

It must be noted that, given a system with nominal parameter $\bar{p}$ and given a prescribed NSR, the estimated value $\hat{p}$ is a stochastic quantity, as well as e_{tot}. Indeed, there are three sources of randomness in the entire procedure:

i) The generation of the output noise sequence v_h.
ii) The initial guess p^0 for the minimization algorithm of problem (4.29). In our experiments, each component p_i^0 of the initial guess has been randomly selected in a large interval containing the nominal value, namely $0.5\bar{p}_i \leq p_i^0 \leq 1.5\bar{p}_i$.
iii) The evolution of the genetic minimization algorithm, which is based on a stochastic generation (and subsequent evaluation) of candidate parameter values.

Therefore, given a system with nominal parameter value $\bar{p}$, for each noise-to-signal ratio NSR the estimation procedure has been replicated 10 times, and the *average* total estimation error has been computed over the 10 replicates.

To illustrate the results obtained with the chemical reaction model (4.8), we firstly recall the system equations:

$$\dot{x}_1 = 0.1544(p_1 + x_3) - x_1 - x_1 x_2^2, \tag{4.32a}$$
$$\dot{x}_2 = p_2(x_1 + x_1 x_2^2 - x_2), \tag{4.32b}$$
$$\dot{x}_3 = p_3(x_2 - x_3). \tag{4.32c}$$

The output variable is $y = x_2$, and the nominal parameter value (supposedly unknown) is $\bar{p} = (\bar{p}_1, \bar{p}_2, \bar{p}_3) = (65, 200, 47)$. System (4.32) has been integrated on $0 \leq t \leq 2.5$, and the output has been sampled with $\tau = 0.0006$ to produce data series with $T = 4166$ points. The noise-free

"Empirical PPP" is very similar to that of Fig. 4.3(a) (which is actually obtained with an infinite sampling frequency).

The results are summarized in Table 4.1, where N, M, and q are, respectively, the half-width of the filtering window ($N = 0$ means no filtering), the half-width of the peak-detection window, and the degree of the interpolating polynomial (see Secs. 4.1 to 4.3).

Table 4.1 Results of the parameter estimation procedure.

NSR (N, M, q)	0 $(0, 1, 10)$	5 $(4, 25, 10)$	10 $(7, 31, 10)$
$\bar{e}_{tot}$	10.75	6.17	12.07

Figure 4.9 displays the results in one of the high-noise experiments (i.e. NSR=10). Although the points of the "Empirical PPP" are scarce and rather dispersed (compare with Fig. 4.3(a)), the accordance between the interpolating polynomials F and $\hat{F}_{\hat{p}}$ is quite satisfactory. To summarize, the accuracy of the parameter estimation method turned out to be sufficiently good (on average) in the considered example. Even better results have been obtained on other chaotic systems [Piccardi (2006)].

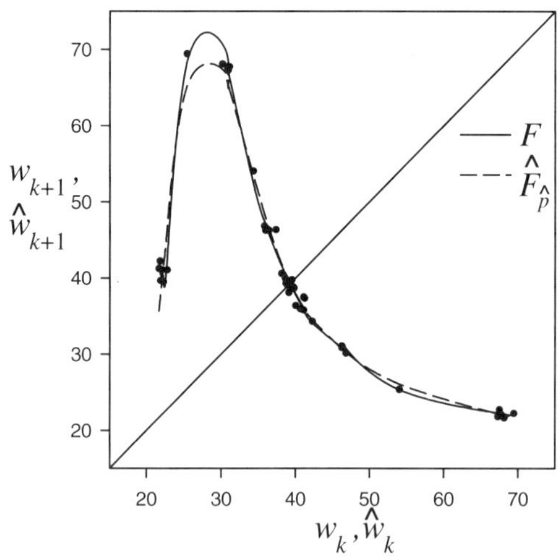

Fig. 4.9 An "Empirical PPP" (NSR=10) with its interpolating polynomial F, and the interpolating polynomial $\hat{F}_{\hat{p}}$ of the "Model PPP".

4.5 Concluding Remarks

Modeling a dynamical system by means of its Peak-to-Peak Dynamics is rather straightforward and yields a very compact model, that can be effectively exploited for control system design and for parameter estimation. Simple filtering techniques allows one to cope with the data corruption due to the noise and to the finite sampling frequency.

For brevity, the exposition could not touch upon some other interesting aspects of the method. For example, it can easily be ascertained that, when Peak-to-Peak Dynamics exist, not only the amplitude of the next peak but also its time of occurrence can be accurately predicted [Candaten (2000)]. This extra-information allows one to enrich the scope of applicability of the method, for example by formulating control problems with different goals, e.g. min-max problems involving return times or mixed amplitude-return times indicators [Piccardi (2000)].

It is worth noticing that peak-to-peak analysis is essentially a numerical, model-free method. In fact, in order to determine the reduced model, one needs simulations or experiments on the continuous-time system: the reduced model is then identified by means of a suitably organized fitting procedure. As a consequence, the method can be extended to classes of systems other than those considered in the paper. For example, interesting results have been obtained with delay-differential systems [Piccardi (2001)], a specific class of infinite-dimensional systems, and with networks of various topologies [Maggi (2006)].

References

Aguirre L.A. (2000). Black and grey-box modeling of nonlinear systems: identification and analysis from time series, in Controlling Chaos and Bifurcations in Engineering Systems, ed. Chen, G. (CRC Press, Boca Raton, FL) pp. 23-43.

Abarbanel H.D.I., Korzinov L., Mees A.I. and Rulkov N.F. (1997). Small force control of nonlinear systems to given orbits, *IEEE Trans. Circuits and Systems I* 44, pp. 1018-1023.

Alligood K.T., Sauer T.D. and Yorke J.A. (1996). Chaos, An Introduction to Dynamical Systems (Springer, New York).

Bertsekas D.P. (1995) Dynamic Programming and Optimal Control (Athena Scientific, Belmont, Mass.).

Baake E., Baake M., Bock H.G., and Briggs K.M. (1992). Fitting ordinary differential equations to chaotic data, *Phys. Rev. A* 45, pp. 5524-5529.

Candaten M. and Rinaldi S. (2000). Peak-to-peak dynamics: a critical survey, *Int. J. Bifurcation and Chaos* 10, pp. 1805-1820.

Eckmann J.P. and Ruelle, D. (1985). Ergodic theory of chaos and strange attractors, *Rev. Mod. Phys.* 57, 617-656.
Goldberg D.E. (1989). Genetic Algorithm in Search Optimization and Machine Learning (Addison-Wesley, Reading, MA).
Gen M. and Cheng R. (2000). Genetic Algorithms and Engineering Optimization (Wiley, New York).
Guckenheimer J. and Holmes P. (1983) Nonlinear Oscillations, Dynamical Systems, and Bifurcations of Vector Fields (Springer, New York).
Julier S.J. and Uhlmann J.K. (2004). Unscented filtering and nonlinear estimation, *Proc. IEEE* 92, pp. 401-422.
Lorenz E.N. (1963). Deterministic nonperiodic flow, *J. Atmospheric Science* 20, 130-141.
Mees A.I. (2000). Reconstructing input-output dynamics from time series, in *Controlling Chaos and Bifurcations in Engineering Systems*, ed. Chen, G. (CRC Press, Boca Raton, FL) pp. 1-21.
Maggi S. and Rinaldi S. (2006). Synchronization and peak-to-peak dynamics in networks of low dimensional chaotic oscillators, *Int. J. Bifurcation and Chaos*, in press.
Obinata G. and Anderson B.D.O. (2001). Model Reduction for Control System Design (Springer, New York).
Piccardi C. (2001). Controlling chaotic oscillations in delay-differential systems via peak-to-peak maps, *IEEE Trans. Circuits and Systems I* 48, pp. 1032-1037.
Piccardi C. (2006). Parameter estimation for systems with low-dimensional chaos, in preparation.
Peng B., Scott S.K. and Showalter K. (1990). Period doubling and chaos in a three variable autocatalator, *J. Phys. Chem.* 94, pp. 5243-5246.
Peng B., Petrov V. and Showalter K. (1991). Controlling chemical chaos, *J. Phys. Chem.* 95, pp. 4957-4959.
Piccardi C. and Rinaldi S. (2000). Optimal control of chaotic systems via peak-to-peak maps, *Physica D* 144, pp. 298-308.
Piccardi C. and Rinaldi S. (2002). Control of complex peak-to-peak dynamics, *Int. J. Bifurcation and Chaos* 12, pp. 2927-2936.
Piccardi C. and Rinaldi S. (2003). The impact of noise and sampling frequency on the control of peak-to-peak dynamics, *Int. J. Bifurcation and Chaos* 13, pp. 1579-1586.
Pivka L., Wu C.W. and Huang A. (1996). Lorenz equation and Chua's equation, *Int. J. Bifurcat. Chaos* 6, pp. 2443-2489.
Rössler O.E. (1979). An equation for hyperchaos, *Phys. Lett. A* 71, pp. 155-157.
Strogatz S.H. (1994). Nonlinear Dynamics and Chaos (Addison Wesley, Reading, MA).
Voss H.U., Timmer J. and Kurths J. (2004). Nonlinear dynamical system identification from uncertain and indirect measurements, *Int. J. Bifurcation and Chaos* 14, pp. 1905-1933.

Chapter 5

Synchronization of Complex Networks

S. Boccaletti and M. Chavez

5.1 Introduction

During the last decades the emergence of collective dynamics in large networks of coupled units has been investigated in many different areas of science. In particular, the effect of synchronization in systems of coupled oscillators nowadays provides a unifying framework for phenomena arising in fields such as optics, chemistry, biology and ecology (for recent reviews see [Pikovsky (2001); Boccaletti (2002); Manrubia (2004)]). Recently, complex networks have provided a challenging framework for the study of synchronization of dynamical units, based on the interplay between complexity in the overall topology and local dynamical properties of the coupled units [Strogatz (2001); Boccaletti (2006)].

This Chapter aims at reviewing the main techniques that have been proposed for assessing the propensity for synchronization (synchronizability) of a given networked system. We will describe the main applications, especially in the view of selecting the optimal topology in the coupling configuration that provides enhancement of the synchronization features.

5.2 Synchronization of Interacting Oscillators

Synchronization is a process wherein many systems (either equivalent or non equivalent) adjust a given property of their motion due to a suitable coupling configuration, or to an external forcing. Historically, synchronization phenomena have been actively investigated since the earlier days of physics. Already in the 17th century, indeed, Christian Huygens discovered that two pendulum clocks hanging at the same beam were able to perfectly synchronize their phase oscillations [Hugenii (1673)]. Other early found examples of synchronous motion include the synchronized flashing of fireflies, the collective chirping of crickets, as well as the phase locking of running and breathing in mammals [Pikovsky (2001); Boccaletti (2002); Manrubia (2004)].

Initially, the attention was mainly devoted to synchronization of periodic systems, while recently the search for synchronization has moved to chaotic systems [Boccaletti (2002)].

When chaotic elements are coupled, many different synchronization phenomena can take place, from complete or identical synchronization [Fujisaka (1983); Afraimovich (1986); Pecora (1990)], to phase [Rosenblum (1996); Rosa (1998)] and lag synchronization [Rosenblum (1997)].

Generalized synchronization is described [Rulkov (1995); Kocarev (1996)], intermittent lag synchronization in [Rosenblum (1997); Boccaletti (2000)], imperfect phase synchronization [Zaks (1999)], and almost synchronization in [Femat (1999)].

Complete synchronization (or identical synchronization) is the simplest form of synchronization, and consists in a perfect matching of the trajectories of identical chaotic systems in the course of the time. Generalized synchronization, instead, considers different systems and associates the output of one system to a given function of the output of another system [Rulkov (1995); Kocarev (1996)]. Coupled non identical oscillators can also reach the phase synchronization regime, where the suitably defined phases of the oscillators become locked, while the amplitudes remain chaotic and generally uncorrelated [Rosenblum (1996)]. Lag synchronization implies that the trajectories of two coupled oscillators shifted by a lag time τ_{lag} become identical [Rosenblum (1997)]. This latter phenomenon can also occur intermittently, producing the intermittent lag synchronization, where the coupled systems are most of the time verifying the condition for lag synchronization, but persistent bursts of local non synchronous behavior may intermittently affect their dynamics [Rosenblum (1997); Boccaletti (2000)]. Analogously, imperfect phase synchronization is the regime where intermittent phase slips take place within a phase synchronization process [Zaks (1999)]. Finally, in the almost synchronization regime, the difference between a subset of the variables of two systems becomes asymptotically bounded [Femat (1999)].

5.3 From Local to Long-Range Connections

In the study of interacting oscillators, three types of coupling schemes are currently considered: global coupling where each unit interacts with all the others, local coupling where an element interacts with its neighbors (defined by a given metric), and non-local or intermediate couplings. In an ensemble of globally coupled oscillators, the structure of connections is not dependent on the spatial distance between the oscillators. In such networks, each individual unit is influenced by the global dynamics through an interaction of a *mean-field* type. In networks of both limit-cycle and chaotic oscillators with slightly different oscillation modes, a phase transition associated to a collective and coherent behavior may be observed if the coupling strength is increased [Kuramoto (1984); Winfree (1967); Pikovsky (1996); Kiss (2002)].

In an opposite approach, several studies considered oscillators embedded in D-dimensional lattices, where each unit interacts only with its nearest neighbors. However, this approach presents severe theoretical difficulties for its analysis [Strogatz (2000, 1988); Matthews (1991)]. Although different collective regimes can be observed (global or partial synchronization, anti-phase synchronization, phase clustering, and coherence resonance) in large ensembles of coupled chaotic or periodic elements, cooperative dynamics are strongly dependent on the size and dimension of the lattices, as well as on the distribution of the eigenfrequencies of each oscillator [Sakaguchi (1987); Klevecz (1992); Osipov (1997)] [Zhao (1998); Belykh (2000); Zhang (2001); Liu (2001); Zhou (2002)].

Inspired from some biological systems, where the cell-to-cell interaction is mediated by rapidly diffusive chemical transmitters, some non-local coupling schemes have been proposed to take into account the natural decay of the information content with the distance. In both theoretical and experimental studies, the dependence of the spatial correlation on the range of non-local coupling was found to decay with a power law [Kuramoto (1995, 1996, 1997)] [Rogister (2004)]. Other long-range interaction schemes include the power-law couplings, in which the interaction strength decays with the lattice distance following a power law $r^{-\alpha}$ where r is the distance between the oscillators and $\alpha \geq 0$ defines the range of the coupling: the limit $\alpha \to 0$ corresponds to the nearest neighbor coupling, whereas an uniform coupling is obtained for $\alpha = 0$. In large networks of limit-cycle or chaotic oscillators, power law coupling schemes were found to yield collective behaviors which also depend on the range of coupling (given by the parameter α) and the distribution of the modes (natural frequencies) of each unit [Rogers (1996); Pinto (2002); Maródi (2002); Massunaga (2002)].

In real physical systems consisting of interacting units, a certain degree of randomness may exist not only in the intrinsic properties of each unit, but also in the connections between them. Several studies have shown that this heterogeneity of coupling weights may strongly affect the collective behaviors in either global or local coupling networks. Hence, the emergence of new features in the cooperative dynamics may be expected for disordered connections.

In a pioneering work [Niebur (1991)], phase locking in an lattice array was studied for different topologies: regular nearest neighbors coupling, random-local coupling (where the weights of connections were randomly drawn from a Gaussian function related to the distance to its nearest neighbors), and long-range sparse connections (where the probability of connecting a node to a given number of neighbors was ruled by a Gaussian function centered on the current node). For the same overall coupling strength, long-range interactions were found to produce a faster and more robust synchronization than the local coupling topologies (regular or random).

5.4 The Master Stability Function

In the study of synchronization, a very relevant problem is to assess conditions for the stability of the synchronous behavior for a generic network topology, and for a generic coupling configuration. The Master Stability Function approach was originally introduced for arrays of coupled oscillators [Pecora (1998a)], and it has been later extended to the case of a complex networks of dynamical systems coupled with arbitrary topologies [Barahona (2002); Chen (2003); Hu (1998); Zhan (2000)].

5.4.1 *The case of continuous time systems*

Let us consider a generic network of N coupled dynamical units, each one of them giving rise to the evolution of a m-dimensional vector field $\mathbf{x}_i$ ruled by a local set of ordinary differential equations $\dot{\mathbf{x}}_i = \mathbf{F}_i(\mathbf{x}_i)$. The equation of motion reads:

$$\dot{\mathbf{x}}_i = \mathbf{F}_i(\mathbf{x}_i) - \sigma \sum_{j=1}^{N} G_{ij}\mathbf{H}[\mathbf{x}_j],\ i = 1, \ldots, N. \tag{5.1}$$

Here $\dot{\mathbf{x}}_i = \mathbf{F}_i(\mathbf{x}_i)$ governs the local dynamics of the ith node $\mathbf{x}_i \in \mathbb{R}^d$, the output function $\mathbf{H}[\mathbf{x}]$ is a vectorial function, σ is the coupling strength, and G is a coupling matrix, accounting for the topology of the network wiring.

In order to proceed with the analytic treatment, we will make the following explicit assumptions:

(i) the network is made of *identical* ($\mathbf{F}_i(\mathbf{x}_i) = \mathbf{F}(\mathbf{x}_i)$, $\forall i$) and *autonomous* systems, i.e. the local dynamics $\mathbf{F}$ and the output function $\mathbf{H}$ do explicitly depend neither on the node index i nor on time.

(ii) the coupling matrix G_{ij} is a real zero-row sum matrix, i.e. the diagonal elements are given by $G_{ii} = -\sum_{j \neq i} G_{ij}$. The matrix G is assumed to be diagonalizable, and therefore there exists a set of eigenvalues λ_i (of associated orthonormal eigenvectors $[\mathbf{v}_i]$), such that $G\mathbf{v}_i = \lambda_i \mathbf{v}_i$ and $\mathbf{v}_j^T \cdot \mathbf{v}_i = \delta_{ij}$. Further, because the zero row condition, the spectrum of eigenvalues is entirely semi-positive, i.e. $\lambda_i \geq 0 \ \forall i$.

(iii) The output function $\mathbf{H}[\mathbf{x}]$ is a linear function, i.e. units in the network are linearly coupled. In practice, $\mathbf{H}[\mathbf{x}]$ simply projects on a subspace of $\mathbb{R}^d$ for example by extracting one variable. Even though the general ideas are not limited to linear coupling configurations, this assumption will help us in clarifying the general method that can be used also for non linear coupling schemes.

Taken together, these assumptions ensure that the coupling term in Eq. (5.1) vanishes, making the synchronized state an invariant manifold. The synchronized state $\mathbf{x}_i(t) = \mathbf{x}_s(t), \forall i$ with $\dot{\mathbf{x}}_s = \mathbf{F}(\mathbf{x}_s)$ is a -possibly unstable- solution of Eq. (5.1), and therefore stability of the synchronous state reduces to take care of the system's dynamical properties along all directions in phase space that are *transverse* to the synchronization manifold.

A major advancement in the linear stability analysis of Eq. 5.1 for the synchronized state was the observation that the analysis can be divided into a topological part and a dynamical part [Pecora (1998a)]. The topological part involves the calculation of the spectrum of the coupling matrix G, while for the dynamical part the Master Stability Function for $\mathbf{F}$ and $\mathbf{H}$ needs to be obtained.

A necessary condition for stability of the synchronization manifold [Pikovsky (2001); Boccaletti (2002)] is that the set of $(N-1)*m$ Lyapunov exponents that corresponds to phase space directions transverse to the $m-$dimensional hyperplane $\mathbf{x}_1 = \mathbf{x}_2 = \ldots = \mathbf{x}_N = \mathbf{x}_s$ be entirely made of negative values. Despite being a universal stability standard, it is important to warn that this condition lacks to be sufficient. Lyapunov exponents are asymptotic averages, and, as so, they account for global stability properties. However, they do not warrant that there are not unstable invariant sets in the synchronization manifold itself [Ashwin (1994)], or locally unstable areas of the attractor [Gauthier (1996)] that can possibly produce bubbling or bursting of the dynamics away from the synchronous state (when, e.g., noise is acting on the system).

Let $\delta\mathbf{x}_i(t) = \mathbf{x}_i(t) - \mathbf{x}_s(t) = (\delta x_{i,1}(t), \ldots, \delta x_{i,m}(t))$ be the deviation of the ith vector state from the synchronization manifold, and consider the $m \times N$ column vectors $\mathbf{X} = (\mathbf{x}_1, \mathbf{x}_2, \ldots, \mathbf{x}_N)^T$ and $\delta\mathbf{X} = (\delta\mathbf{x}_1, \ldots, \delta\mathbf{x}_N)^T$.

The equation of motion for the perturbation $\delta\mathbf{X}$ can be straightforwardly obtained by expanding the Eq.(5.1) in Taylor series of 1st order around the synchronized state, which gives

$$\delta\dot{\mathbf{X}} = \left[\mathbb{I}_N \otimes \mathbf{JF}(\mathbf{x}_s) - \sigma G \otimes \mathbf{JH}(\mathbf{x}_s)\right]\delta\mathbf{X}, \tag{5.2}$$

where $\otimes$ stands for the direct product between matrices, and J denotes the Jacobian operator.

Solving this system, referred as the variational equations, is rather complicated because its (possible) high dimensionality. Instead, one can notice that the arbitrary state $\delta\mathbf{X}$ can be written as $\delta\mathbf{X} = \sum_{i=1}^{N} \mathbf{v}_i \otimes \zeta_i(t)$ $[\zeta_i(t) = (\zeta_{1,i}, \ldots, \zeta_{m,i})]$. By applying $\mathbf{v}_j^T$ to the left side of each term in Eq. (5.2), one finally obtains a set of N blocks (the variational equations in an eigenmode form) for the coefficients $\zeta_i(t)$ that read

$$\dot{\zeta}_j = \mathbf{K}_j \zeta_j, \tag{5.3}$$

where $j = 1, \ldots, N$ and $\mathbf{K}_j = \left[\mathbf{JF}(\mathbf{x}_s) - \sigma\lambda_j \mathbf{JH}(\mathbf{x}_s)\right]$ is the evolution kernel.

It is important to notice that each block in Eq. (5.3) corresponds to a set of m conditional Lyapunov exponents (the kernels $\mathbf{K}_j$ are calculated on the synchronization manifold) along the eigenmode corresponding to the specific eigenvalue λ_j. Because the zero-row condition of the coupling matrix, $\lambda_1 \equiv 0$ with associated eigenvector $\mathbf{v}_1 = \frac{\pm 1}{\sqrt{N}}\{1, 1, \ldots, 1\}^T$ entirely defines the synchronization manifold. All the other eigenvalues λ_i ($i = 2, \ldots, N$) of associated eigenvectors $\mathbf{v}_i$ span all the other directions of the $m \times N$-dimensional phase space transverse to the synchronization manifold.

A necessary condition for the linear stability of the synchronized state is therefore that all Lyapunov exponents associated with λ_i for each

$i = 2, \ldots, N$ (the directions transverse to the synchronization manifold) are negative. It should be stressed that this criterion only guarantees that the system remains synchronized. As the eigenmode associated to the eigenvalue $\lambda_1 = 0$ lies entirely within the synchronization manifold, the corresponding m conditional Lyapunov exponents equal those of the single uncoupled system $\dot{\mathbf{x}} = \mathbf{F}(\mathbf{x})$. The synchronized state itself can well have a positive Lyapunov exponent and be chaotic.

For an arbitrary coupling matrix, the eigenvalues spectrum is either real or made of pairs of complex conjugates, and the terms $\sigma\lambda_i$ may take, in general, complex values [Cvetkovic (1995)]. We can therefore consider the following d-dimensional parametric equation

$$\dot{\zeta} = \mathbf{JF}(\mathbf{x}_s)\zeta - (\alpha + i\beta)\mathbf{H}[\zeta], \tag{5.4}$$

that, once coupled with the d-dimensional local nonlinear evolution of the networked system ($\dot{\mathbf{x}}_s = \mathbf{F}(\mathbf{x}_s)$), gives a set of d Lyapunov exponents for any choice of α and β. The surface ($\Lambda(\alpha,\ \beta)$) in the parameter plane tracing the maximum of such exponents is called Master Stability Function [Pecora (1998a)], and fully defines the stability properties of the synchronization manifold. Given a coupling matrix, the stability of the different eigenmodes can be evaluated by the sign of $\Lambda(\alpha,\ \beta)$. Varying real coupling terms in $\alpha + i\beta$ can be interpreted as having an effect of damping the transverse perturbations of the synchronization manifold, whereas changes in imaginary terms have a rotation effect between different modes [Fink (2000)]. If all the eigenmodes (associated with the different $\lambda_i,\ i \geq 2$) are stable ($\Lambda < 0$) the synchronous state is then stable.

To illustrate the above arguments, let us to consider an example of a network of coupled chaotic Rössler oscillators [Rössler (1976)]. The dynamics is ruled by Eq. (5.1), with $\mathbf{x} = (x, y, z)$, $\mathbf{F}(\mathbf{x}) = \mathbf{F}(\mathbf{x}) = [-y - z,\ x + 0.165y,\ 0.2 + z(x - 10)]$. The Master Stability Function is depicted in Fig. (5.1) in the complex plane $(\alpha, i\beta)$ for two type of couplings: $\mathbf{H}[\mathbf{x}] = y$ and $\mathbf{H}[\mathbf{x}] = x$.

For both the x and y–couplings $\Lambda(0, 0) > 0$ because this is just the case of uncoupled chaotic systems. For the y-coupling (Fig. 5.1(a)), we observe that with increasing α the Master Stability Function $\Lambda(\alpha,\ \beta)$ drops below zero and remains negative even for large values of coupling. Therefore a sufficiently large coupling guarantees in this case the stability of the synchronized state. For the x-coupling (Fig. 5.1(b)) however, $\Lambda(\alpha,\ \beta)$ is only negative within a finite region and stability is lost for large couplings strengths. In both examples of Fig. 5.1, the regions of stability $\Lambda(\alpha,\ \beta) < 0$

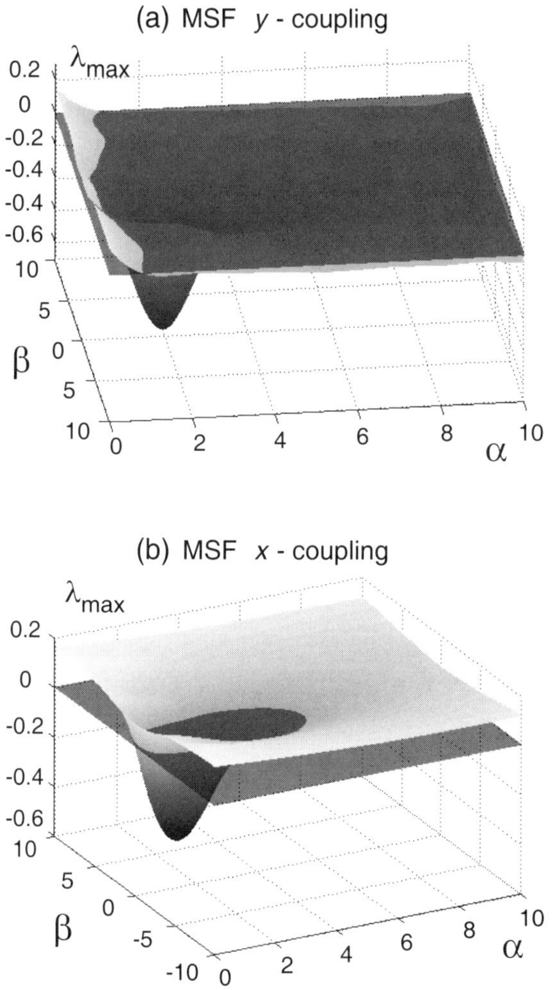

Fig. 5.1 The Master Stability Function (MSF) in the parameter space (α, β) for a network of chaotic Rössler oscillators coupled through a y-coupling (a) and x-coupling (b) schemes. The regions of stability $\Lambda(\alpha,\ \beta) < 0$ are given by the surfaces under the dark plane at $\Lambda = 0$.

are given by the surfaces under the plane at $\Lambda = 0$. Generically, there exists a bounded region of stability in the complex plane (symmetric in the imaginary values about the real axis) where $\Lambda(\alpha, \beta) < 0$ and an arbitrary increasing of the coupling strength may induce a destabilization of the synchronous state [Pecora (1998a); Fink (2000); Barahona (2002)].

5.4.2 *The Master stability function for coupled maps*

All the results presented above can be straightforwardly extended to arbitrary arrays of coupled maps [Fink (2000)]. Let us consider a network of N coupled maps, whose dynamical evolution is ruled by

$$\mathbf{x}_{t+1}^i = \mathbf{f}(\mathbf{x}_t^i) - \sigma \textstyle\sum_{j=1}^N G_{ij}\mathbf{H}(\mathbf{x}_t^j),\ i = 1, \ldots, N. \tag{5.5}$$

were $\mathbf{x}_{t+1}^j = \mathbf{f}(\mathbf{x}_t^i)$ governs the local dynamics of each map, and the output function $\mathbf{H}[\mathbf{x}]$ defines the coupling function between the nodes. As for the time continuous case, σ and G define the coupling strength and the coupling matrix (with eigenvalues $0 = \lambda_1 \leq \lambda_2 \leq \ldots \leq \lambda_N$), respectively.

Let us consider $\mathbf{x}_t^i = \mathbf{x}_t^s, \forall\ i$ to be the synchronized state of coupled variables. The linear stability of the synchronization manifold is governed by the variational equations of Eq. (5.5) which have the following block diagonalized form

$$\eta_{t+1}^i = \Big[\exp\big[\Lambda(\sigma\lambda_i)\big]\Big]\eta_t^i = \Big[\mathbf{Jf}(\mathbf{x}_t^s) - \sigma\lambda_i\mathbf{JH}(\mathbf{x}_n^s)\Big]\eta_t^i, \tag{5.6}$$

where $\Lambda(\epsilon\lambda_i)$ is the Lyapunov exponent.

For illustration purposes (the stability of coupled map networks are studied in detail by [Fink (2000); Gade (2000)] [Chen (2003); Rangarajan (2002); Li (2003); Lind (22004); Lu (2004c)] [Stefański (2004); Gong (2005)]), let us consider the coupling function $\mathbf{H}(\mathbf{x}) = \mathbf{f}(\mathbf{x})$. Under this assumption, $\mathbf{JH} = \mathbf{Jf}$ and the system Eq. (5.6) reduces to

$$\eta_{t+1}^i = \big[\mathbf{Jf}(\mathbf{x}_t^s) - \sigma\lambda_i\mathbf{Jf}(\mathbf{x}_t^s)\big]\eta_t^i, \tag{5.7}$$

It turns out that he explicit conditions of stability for chaotic maps can be obtained as a function of eigenvalues λ_i and the local map properties:

$$\frac{1 - \exp(-\mu_{max})}{\lambda_2} < \sigma < \frac{1 + \exp(-\mu_{max})}{\lambda_N}, \tag{5.8}$$

where μ_{max} is the maximal Lyapunov exponent of the uncoupled local map. In order to achieve a stable synchronization, the above inequality must be satisfied for all eigenvalues λ_i of the matrix G, except one $\lambda_1 = 0$, which corresponds to the motion within the synchronization manifold.

5.5 Key Elements for the Assessing of Synchronizability

Let us consider the case where the coupling matrix G has a real spectrum (as e.g. for symmetric coupling). For a large class of oscillatory systems, the Master Stability Function is negative in a finite parameter interval (α_1, α_2) [Pecora (1998a); Fink (2000); Barahona (2002)]. From Eq. (5.4) it is clear that the stability condition can only be satisfied for some σ when $\lambda_N/\lambda_2 < \alpha_2/\alpha_1$. As the ratio α_2/α_1 depends on $\mathbf{F}(\mathbf{x})$ and $\mathbf{H}[\mathbf{x}]$, the key quantity for assessing synchronizability of a network is the eigenratio λ_N/λ_2 which only depends on the coupling matrix G: the smaller λ_N/λ_2 is (the more packed the eigenvalues of G are), the higher is the chance of having all Lyapunov exponents in the stability range for some coupling strength σ [Barahona (2002)].

For an arbitrary coupling matrix G with real values, the spectrum is in general complex ($\lambda_1 = 0$; $\lambda_k = \lambda_k^r + j\lambda_k^i$, $k = 2, \ldots, N$). The stability condition now requires that for a given σ, the set $\{\sigma\lambda_k,\ k = 2, \ldots, N\}$ must be entirely contained in the region R where the Master Stability Function is negative $\Lambda(\alpha, \beta) < 0$. Since the boundary of this region can in principle be complicated, it is not possible to provide simple universal stability conditions.

The Gerschgorin disk theorem [Gerschgorin (1931)] asserts that each eigenvalue of G lies in at least one of the disks in the complex plane, defined by

$$\Gamma_i = \left\{ z : \mid z - d_i \mid \leq \sum_{j \neq i} \mid G_{ij} \mid \right\}, \quad i = 1, \ldots, N \tag{5.9}$$

where d_i are the diagonal elements of G. The spectrum of G is then fully contained within the union of those disks ($\{\lambda_l\} \subset \cup_i \Gamma_i$). Notice that if the coupling matrix is made of real entries, in all cases the circles have centers lying on the real axis.

In addition to the assumptions made previously, i.e. G is a zero row-sum matrix and $d_i = \sum_{j \neq i} \mid G_{ij} \mid$ (all non zero off diagonal elements are negative), let us suppose the diagonal elements of the coupling matrix G to be properly normalized to $G_{ii} = 1$. Under these assumptions, the Gerschgorin disk theorem implies that all eigenvalues are fully contained within the unit circle centered at 1 on the real axis ($\mid \lambda_k - 1 \mid \leq 1$, $k = 2, \ldots, N$). This yields the following inequalities: *i)* $0 < \lambda_2^r \leq \ldots \leq \lambda_N^r \leq 2$ (the multiplicity of $\lambda_i = 0$ corresponds to the number of separated connected subnetworks produced by the coupling matrix [Cvetkovic (1995)]), and *ii)* $\mid \lambda_l^i \mid \leq 1$, $\forall l$.

These properties warrant that the largest eigenvalue λ_N will never diverge, independently of the network size N, and regardless on the local dynamics $\mathbf{F}$. This is essential to provide a consistent and unique mathematical framework within which one can formally assess the relative merit of one topology against another for optimizing the propensity for synchronization of a network, *regardless* on the specific properties of the local dynamics.

By calling $\mathcal{R}$ the bounded region in the complex plane where the master stability function $\Lambda(\nu)$ provides a negative Lyapunov exponent, the stability condition for the synchronous state is that the set $\{\sigma\lambda_l, l = 2, \ldots, N\}$ be entirely contained in $\mathcal{R}$ for a given σ. This is best accomplished for connection topologies that simultaneously make the ratio $\frac{\lambda_N^r}{\lambda_2^r}$ and $\mathcal{M} \equiv \max_{l\geq 2}\{\mid \lambda_l^i \mid\}$ as small as possible. A network architecture which makes such quantities as small as possible provides therefore very good conditions for synchronization [Hwang (2005)].

In previous works [Wu (1995a,b, 1996)], the value of the coupling strength necessary to synchronize a network of arbitrarily coupled identical systems was proposed to be ruled by the structure of the coupling matrix, and by the coupling strength necessary to synchronize two elements of the network. The so-called Wu-Chua conjecture implies that the coupling threshold for synchronization in such an array is $\sigma_c = \frac{2\sigma_2}{|\lambda_2|}$ where σ_2 is the synchronization threshold for synchronization of two oscillators.

We stress that this conjecture is only valid under the assumption that the stability of the transversal mode associated with λ_2 induces the stability of all other modes. This implies that a network of coupled identical oscillators can always be synchronized for sufficiently high coupling strengths. While the analysis of networks formed by coupled Lorenz or Chua's oscillators has confirmed this conjecture, many other relevant cases (see the examples in Fig. 5.1) exist where an increasing of the coupling strength results in a destabilization of the synchronous state.

5.5.1 *Bounding the eigenratio* $\frac{\lambda_N}{\lambda_2}$

Previous studies showed that the stability of the synchronization manifold is sensitive to different topological properties such as the heterogeneity of the degree distribution, the average path length and the betweenness centrality [Nishikawa (2003); Hong (2004); Zhao (2005)]. In recent works [Nishikawa (2003); Kocarev (2005)], bounds for the eigenratio of networks with different topologies have been derived based on graph theory concepts. The obtained expressions depend solely on graph theoretic quantities such as the degree distribution, the shortest path length, diameter of the coupling graph, or the number of shortest path lengths passing through the edges (edge's load).

These works allow relating the stability of the synchronization manifold to purely topological properties. Results suggest for instance, that unweighted and symmetrical networks with smaller average path length and smaller betweenness centrality will lead to better synchronizability, though increasing the heterogeneity of the degree distribution will destabilize the synchronous state.

In the case of more general weighted networks, very precise bounds for the eigenratio have been recently provided by [Motter (2005a,b)] [Zhou (2006)]. Such quantities depend solely on the heterogeneity of the so-called nodes intensities, defined as the input weight sums for each node. Under the extra assumptions of an uncorrelated distribution of nodes degrees and non-normalized couplings, the synchronizability of an arbitrary network (with a sufficiently random wiring structure) was found to be increased by decreasing the heterogeneity of the nodes intensities.

5.5.2 *Other approaches for assessing synchronizability*

Master stability function arguments are currently used as a challenging framework for the study of synchronized behaviors in complex networks, especially for understanding the interplay between complexity in the overall topology and local dynamical properties of the coupled units. It is important to notice that, though the stability of synchronization is currently assessed by means of the Lyapunov exponents, a Master Stability Function can also be produced for any other stability criteria that depends on the variational equation of the system [Pecora (1998a)]. Alternatives approaches have been recently developed to assess conditions for synchronization in arbitrary networks of identically coupled units.

An example is the *connection graph stability* method [Belykh (2004a)] [Belykh (2004b, 2005a,b)] that combines Lyapunov functions with graph theory, leading to the determination of a bound for the minimum coupling strength needed to achieve complete synchronization. This approach provides a rigorous condition for synchronization that is based solely on topological properties, which allows also for describing the cases where the coupling matrix G is time-dependent, i.e. $G = G(t)$.

5.6 Synchronizability of Weighted Networks

A basic assumption characterizing most of the early works on synchronization in complex networks is that the local units are symmetrically coupled with uniform undirected coupling strengths (unweighted links). Nevertheless, this simplification does not retain in general the full information on the structure of real networks. There are, indeed, paradigmatic cases where a weighting in the connections has relevant consequences in determining the network's dynamics. In ecology, for instance, the non uniform weight in prey-predator interactions may play a crucial role in determining the food web dynamics [McCann (1998); Berlow (1999)]. In transportation networks, critical quantities that affect the global dynamics are the traffic load of a road, or the number of passengers in subways lines or airports [Latora (2001); Barrat (2004)]. Again, the natural differences of neurons and their dendritic connections may result in distinct capabilities of transmission and information processing in neural networks [Buzsáki (2004); Sporns (2004)].

5.6.1 *Coupling matrices with a real spectra*

A great interest has born on showing how asymmetric weighted coupling configurations may enhance synchronization of complex networks. Recent studies have revealed the strong influence of weighted and asymmetric coupling configurations on the emergence of coherent global behavior. In an initial work [Motter (2005a)], the authors proposed a weighting approach based on degree of the node k_i. Namely, the connections between nodes are weighted by the node degrees such that the coupling term of Eq. (5.1) becomes

$$\frac{\sigma}{k_i^{\beta}} \sum_{j=1}^{N} L_{ij} \mathbf{H}[\mathbf{x}_j], \tag{5.10}$$

where L_{ij} is the usual Laplacian matrix with diagonal entries $L_{ii} = k_i$ and off-diagonal entries $L_{ij} = -1$ if nodes i and j are connected by a link, and $L_{ij} = 0$ otherwise. Weights of the connections are tuned by the parameter β (with $\beta = 0$ recovering the unweighted and undirected configuration).

Although the resulting coupling matrix G is asymmetric for all $\beta \neq 0$, it can be written as a product $G = DL$, where $D = diag\{k_1^{-\beta}, \ldots, k_N^{-\beta}\}$. Using matrix identities [Cvetkovic (1995)], one can show that the eigenvalue spectrum of G coincides with the spectrum of the matrix $W = D^{1/2} L D^{1/2}$, and therefore is real with nonnegative values.

A large class of networks weighted with this approach have an optimal condition for synchronization (the minimum of the eigenratio as a function of β) at $\beta = 1$ [Motter (2005a)]. It is worth to notice that the condition $\beta = 1$ corresponds to a situation where the input strength of the coupling term is equal for all the nodes (the diagonal elements of the coupling matrix are normalized to 1), while, for $\beta \neq 1$ such elements are not normalized and are, in general, size-dependent.

The relevant results of [Motter (2005a)] is the tremendous improvement on the propensity of synchronization, obtained with a weighting procedure that retains information on the local features of the network (the node degree). Further analysis on directed and weighted networks with criteria based on the overall cost of the coupling has been carried out by the same authors [Motter (2005b); Zhou (2006)].

We can notice that the weighting procedure proposed in Eq. (5.10) uses only locally available information of the network (the node's degree). The problem of how a weighting procedure can enhance the propensity for synchronization has been studied also in the framework of properly normalized coupling schemes making use of the information contained in the overall topology [Chavez (2005)], showing that this can give rise to further improvements in the synchronization properties. It turned out to be advantageous to weight a connection with its *load*. The load ℓ_{ij} of the link connecting nodes i and j quantifies the number of shortest paths that are making use of that link [Goh (2001)]. Precisely, for each pair of nodes i', j' in the network, we count the number $n(i', j')$ of shortest paths connecting them. For each one of such shortest paths, we then add $1/n$ to the load of each link forming it.

It is important to notice that the load distribution retains full information on the network structure of pathways at a global level, since the value of each load ℓ_{ij} can be strongly influenced also by pairs of nodes that may be very far away from either nodes i and j. Furthermore, it is essential to highlight that this weighting procedure is totally different from relying on the information on the node degrees, insofar as nodes with low degrees in the networks may be connected through links with very high loads, whereas nodes with high degrees may have links with very poor loads.

With this approach, the coupling term of Eq. (5.1) reads

$$\frac{\sigma}{\sum_{j\in\mathcal{N}_i} \ell_{ij}^{\alpha}} \sum_{j\in\mathcal{N}_i} \ell_{ij}^{\alpha}\mathbf{H}[\mathbf{x}_i - \mathbf{x}_j], \tag{5.11}$$

where α is a real tunable parameter, and $\mathcal{N}_i$ is the set of neighbors to the ith node.

The coupling matrix has now several properties: first, the diagonal elements of G are always normalized to 1. Second, as for the weighting approach based on the node degrees, G is asymmetric for all α but it can be written as a product $G = BL$, where L is a zero row-sum matrix with off-diagonal entries $L_{ij} = -\ell_{ij}^{\alpha}$, and $B = \mathrm{diag}\{\frac{1}{\sum_{\rm j} \ell_{\rm 1j}^{\alpha}}, \ldots, \frac{1}{\sum_{\rm j} \ell_{\rm Nj}^{\alpha}}\}$. The spectrum of G is therefore the same as that obtained from the matrix $W = B^{1/2} L B^{1/2}$, i.e. real and with nonnegative values. Because G has a zero row-sum and the network is connected, the smallest eigenvalue λ_1 is zero while $0 < \lambda_k \leq 2$ for $k \geq 2$ [Cvetkovic (1995)].

Another important point to be stressed concerns the various limits that the coupling term can assume when varying α. The best condition for synchronization of [Motter (2005a)] is recovered for $\alpha = 0$. The limit $\alpha = +\infty$ ($\alpha = -\infty$) induces a unidirected tree structure in the wiring (a network with N nodes and N directed links), such that only the link with maximum (minimum) load is selected as the incoming link for each node. The resulting network will be either connected or disconnected. In the connected (disconnected) case, the ratio λ_N/λ_2 will be equal to 2 ($+\infty$), thus yielding the optimal (the worst) condition for synchronization.

5.6.2 *Numerical simulations*

With the above weighting approaches, we have evaluated the PFS of the following classes of complex networks:

i) SF networks. The used class of scale-free networks is obtained by a generalization of the preferential attachment growing procedure introduced by [Barabási (1999)]. Namely, starting from $m + 1$ all to all connected nodes, at each time step a new node is added with m links. These m links point to old nodes with probability $p_i = \frac{k_i + B}{\sum_j (k_j + B)}$, where k_i is the degree of the node i, and B is a tunable real parameter, representing the initial attractiveness of each node. The advantage of this procedure is that it allows a selection of the γ exponent of the power law scaling in the degree distribution [Albert (2002); Dorogovtesev (2002); Newman (2003)] ($p(k) \sim k^{-\gamma(B,m)}$) with $\gamma(B, m) = 3 + \frac{B}{m}$ in the thermodynamic ($N \to \infty$) limit. While the average degree is by construction $\langle k \rangle = 2m$ (thus independent on B), the heterogeneity of the degree distribution can be strongly modified by B. Notice that the case $B = 0$ recovers the preferential attachment rule originally introduced by [Barabási (1999)].

ii) SW networks. Small world networks were obtained via the rewiring procedure proposed by [Watts (1998)]. We start from a ring lattice with N nodes connected with their k nearest neighbors, and each edge is rewired at random with probability p. Self and duplicate connections are forbidden. This procedure provides a class of networks with a highly homogeneous degree distribution. The limit of regularity is obtained for $p = 0$, whereas the SW regime lies in the intermediate region $0 < p < 1$. Fully random networks are obtained with $p = 1$, for which all nearest neighbor connections are randomly rewired to other nodes in the network. In all cases, the value of k was chosen such that the average degree is $\langle k \rangle = 2m$.

Fig. 5.2 (a) shows the logarithm of λ_N/λ_2 in the parameter space (α, B) for SF networks. The first crucial observation is that the surface of λ_N/λ_2 has a pronounced minimum at $0 < \tilde{\alpha} \simeq 1$ for all values of B above a given $B_c > 0$. For the case $B_c = 0$, the eigenratio decreases as the parameter α increases. Fig. 5.2 (b) reports the behavior of $\log(\lambda_N/\lambda_2)$ as a function of (α, p), for the SW regime. A first observation is that globally, the ability for synchronization increases with the randomness of connections. For all α random configurations ($p = 1$) provide topologies with higher synchronizability than SW or regular networks. However, the eigenratio always displays a minimum at $\tilde{\alpha} \simeq 1$ in all the range of values $0 \leq p \leq 1$.

For a large class of networks (regular lattices, SW networks, random wirings and SF networks with $B_c > 0$), large values of α have a negative effect in the ability of the network to synchronize. The reason for this is that increasing α above $\tilde{\alpha}$ makes that a strongly directed tree structure dominates the network connectivity. Although this may have the effect of increasing the likelihood of synchronization, it also increase the chance of disconnecting the network, rendering the synchronization impossible. Therefore, a second critical value $\alpha_c > \tilde{\alpha}$ can be expected such that the disconnection mechanism dominates over the tree structure induction.

While our study focused on $m = 2$ (chosen as the minimal value of m that produces loops in the growing SF networks), results for larger m values (not reported here) show the following general changes in the scenarios: increasing m reduces $\log(\lambda_N/\lambda_2)$ for all the networks and diminishes the value of $\tilde{\alpha}$. For the SW configuration, as the randomness of wirings increases, large values of m also make the eigenratio to become insensitive to α. For the SF and regular topologies, however, augmenting m also increases the likelihood of disconnecting the network in the limit $\alpha \to \infty$. As a particular case, $m = 1$ always ensures that SF networks remains connected at $\alpha \to \infty$, regardless of B.

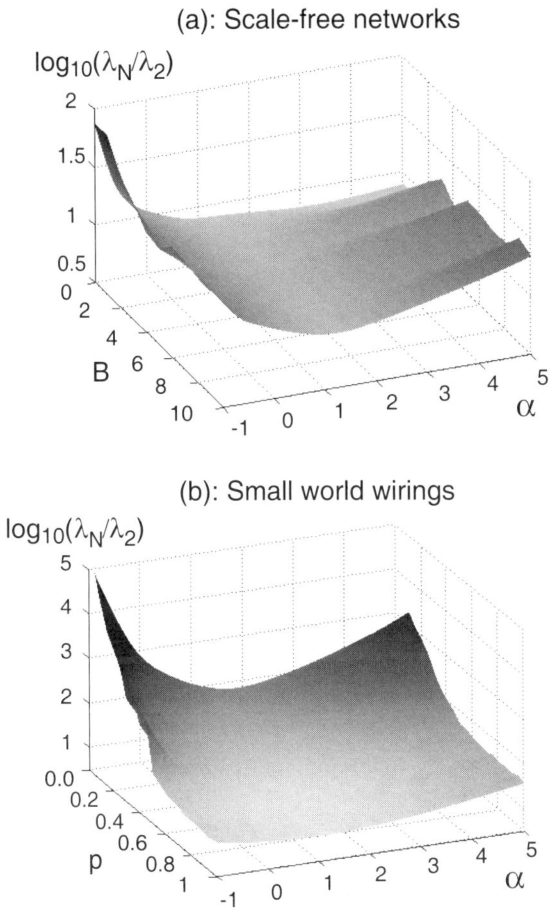

Fig. 5.2 Eigenratio λ_N/λ_2 (in logarithmic scale) *vs.* the dimensionless parameter space (a) (α, B) for SF networks and, (b) (α, p) for SW networks. In all cases $m = 2$, and the reported values refer to an average over 10 realizations of networks with $N = 1,000$ nodes.

5.6.3 *Weighting: local vs global approaches*

We notice that for $\alpha = 0$ one recovers the optimal condition when only the local information on node degrees is used (the condition $\beta = 1$ of [Motter (2005a)]). Because our optimal condition is for $\tilde{\alpha} \simeq 1$ for a large class of networks, this indicates that our weighting procedure based on the link loads *always* enhances the synchronizability of the network. This is reported in Fig. 5.3, where we plot a surface Γ delimiting the area in parameter space where the values of λ_N/λ_2 are smaller than those obtained at $\alpha = 0$. For both the SF and SW networks, one can easily see that for all topological configurations (B or p for the SF or SW networks, respectively) there exists a large parameter interval $\alpha \in [0, \alpha_c]$ with $\Gamma(\alpha) < \Gamma(0)$. In this region the weighting procedure based on the global load information provides a better synchronization propensities than a weighting process based on the local node degrees.

5.6.4 *Coupling matrices with a complex spectra*

Both the approaches developed in [Motter (2005a); Chavez (2005)] were dealing with situations wherein the coupling matrix G had a real spectrum of eigenvalues. Motivated by what happens in social networks [Dorogovtsev (2000); Ramasco (2004)] (where interactions between individuals are not symmetric, and depend upon several social factors, such as age, social class or influence, personal leadership or charisma), [Hwang (2005)] analyzed networks of asymmetrically coupled dynamical units, where asymmetry is explicitly related to an age order among different nodes. In practice, the idea is that the direction of an edge can be determined by an age ordering between the connected nodes. For instance, in growing networks, such age ordering will be naturally related to the appearance order of the node during the growing process.

This is reflected by a zero row-sum coupling matrix G_{ij} in Eq. (5.1) with off diagonal entries:

$$G_{ij} = -A_{ij} \frac{\Theta_{ij}}{\sum_{j \in \mathcal{N}_i} \Theta_{ij}}, \tag{5.12}$$

where A is the adjacency matrix, $\Theta_{ij} = \frac{1-\theta}{2}$ $\left(\Theta_{ij} = \frac{1+\theta}{2}\right)$ for $i > j$ $(i < j)$, and $\mathcal{N}_i$ denotes the set of k_i neighbors of the ith node.

The parameter $-1 < \theta < 1$ governs the coupling asymmetry in the network. Precisely, $\theta = 0$ yields the optimal synchronization condition of [Motter (2005a)], being $\sum_{j \in \mathcal{N}_i} 1 = k_i$, while the limit $\theta \to -1$ $(\theta \to +1)$ gives a unidirectional coupling configuration wherein the older (younger) nodes drive the younger (older) ones.

Though being asymmetric, for $\theta = 0$ the matrix G has a real spectrum of eigenvalues, and the results are the same as those obtained by [Motter (2005a)] for $\beta = 1$. Conversely, for a generic $\theta \neq 0$, the coupling matrix has a spectrum contained in the complex plane ($\lambda_1 = 0$; $\lambda_l = \lambda_l^r + j\lambda_l^i$, $l =$

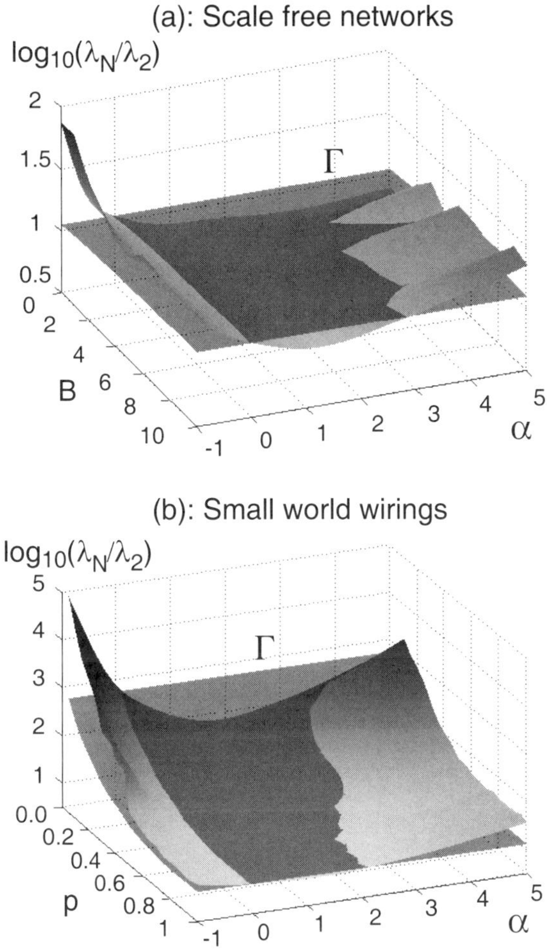

Fig. 5.3 Surfaces Γ obtained at $\alpha = 0$ crossing the surfaces λ_N/λ_2 (in logarithmic scale) in the dimensionless parameter space (a) (α, B) for SF networks and, (b) (α, p) for SW networks.

$2, \ldots, N$). Furthermore, by construction, the diagonal elements of G are normalized to 1 in all possible cases.

As discussed above, if $\mathcal{R}$ is the bounded region in the complex plane where the master stability function provides negative Lyapunov exponents, the stability condition for the synchronous state is that the set $\{\sigma\lambda_l, l = 2, \ldots, N\}$ be entirely contained in $\mathcal{R}$ for a given σ, which is best realized when both the ratio $\frac{\lambda_N^r}{\lambda_2^r}$ and $\mathcal{M} \equiv \max_l\{| \lambda_l^i |\}$ are simultaneously made as small as possible.

With this in mind, [Hwang (2005)] analyzed the effects of heterogeneity in the node degree distribution, by comparing the propensity for synchronization of the class of scale free networks explained above, with that of a highly homogeneous Erdös-Rényi random network [Erdös (1959)], having connection probability $P = \frac{2m}{N-1}$ (giving the same average degree $\langle k \rangle = 2m$), and an arbitrary initial age ordering.

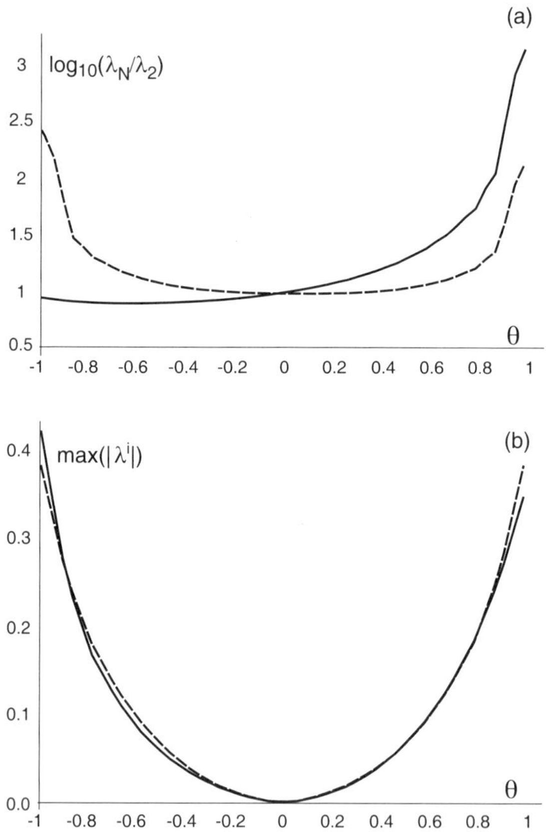

Fig. 5.4 (a) Eigenratio λ_N/λ_2 (in logarithmic scale) and (b) $\mathcal{M}$ *vs.* the tuning parameter θ for SF networks (solid line) and random networks (dashed line). In all cases the reported values refer to an average over 10 realizations of networks with $N = 1,000$ nodes.

Fig. 5.4-(a) reports the behavior of the eigenratio $\frac{\lambda_N^r}{\lambda_2^r}$ *vs.* θ for scale free networks (solid line) and for the random topology (dashed line). It is possible to notice that while the best synchronizability condition for random networks is $\theta = 0$, scale free networks show a better (worse) propensity for synchronization for $\theta \to -1$ ($\theta \to 1$). The imaginary part of the spectra $\mathcal{M}$ *vs.* θ illustrated in Fig. 5.4-(b) indicates that only very small differences exist between the scale free and the random network configurations in the whole range of the asymmetry parameter.

An important point illustrated in Fig. 5.5-(a) is that the monotonically decreasing behavior of $\frac{\lambda_N^r}{\lambda_2^r}$ with θ persists for different degree of heterogeneity (tuned by the parameter B), indicating that synchronization is always enhanced in growing scale free networks when θ becomes smaller. Similarly, results depicted in Fig. 5.5-(b) highlight that the contribution to network synchronizability of the imaginary part of the spectra does not depend significantly on the specific degree distribution.

5.6.5 *Essential topological features for synchronizability*

Enhancing the synchronization in weigthed and directed complex networks is the result of simultaneous ingredients: the first one is that the weighting must induce *a dominant interaction from hub to non hub nodes.* This can be easily understood by a simple example: the case of a star network consisting of a single large hub (the center of the star) and several non-hub nodes connected to the hub. When the dominant coupling direction is from the non-hub nodes to the hub node, synchronization is impossible because the hub receives a set of independent inputs from the different non-hub nodes. In the reverse case (when the center drives the periphery of the star) synchronization can be easily achieved. The very same mechanism occurs in the case of the age order. Indeed, for positive (negative) θ values, the dominant coupling direction is from younger (older) to older (younger) nodes. Now, in growing scale free networks the minimal degree of a node is by construction m and older nodes are more likely to display larger degrees than younger ones, so that a negative θ here induces a dominant coupling direction from hubs to non-hub nodes.

A second fundamental ingredient is that *the network contains a structure of connected hubs influencing the other nodes.* This is in general accounted for by a proper normalization in the off diagonal elements of coupling matrix G, assuring that hubs receive an input from a connected node scaling with

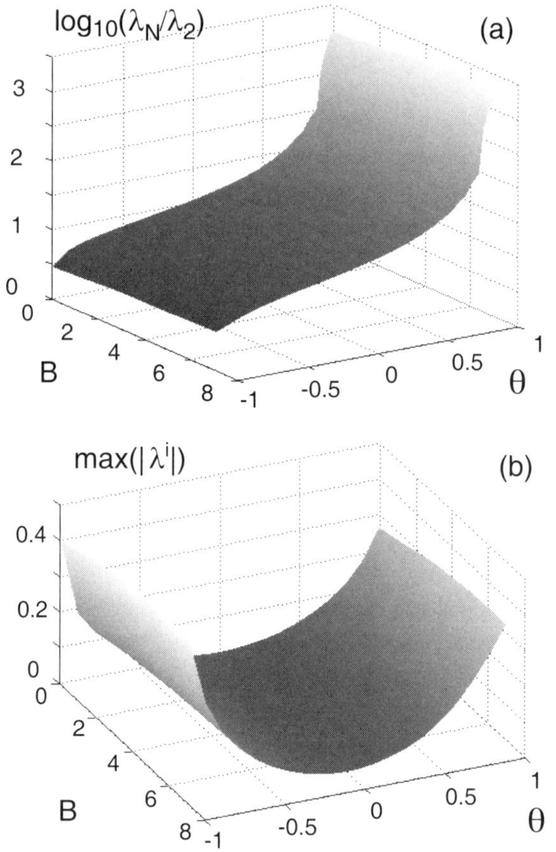

Fig. 5.5 (a) Eigenratio λ_N/λ_2 (in logarithmic scale) and (b) $\mathcal{M}$ *vs.* the dimensionless parameter space (θ, B) for scale free networks.

the inverse of their degree, and therefore the structure of hubs is connected always with the rest of the network in a way that is independent on the network size. For age ordered growing scale free networks, the non zero off diagonal elements of the coupling matrix are given by:

$$\mathcal{C}_{ij} = \frac{1 \mp \theta}{k_i - \theta[k_i - 2\min(i-1, m)]} \quad , \tag{5.13}$$

where $\mp$ stands for $i > j$ and $i < j$, respectively.

The recent works of [Motter (2005a,b)] [Chavez (2005); Hwang (2005); Zhou (2006)] have shown that weighted networks are the most promising framework for the study of how the architectural properties of the connection wiring influence the emergence of a collective (synchronized) behavior. It has to be expected, therefore, that much effort will be stimulated in the forthcoming years to investigate ad-hoc weighted configurations, for assessing more deeply the underlying mechanisms for the formation of collective dynamics in all those circumstances where the weighting features can be directly extracted from available data.

5.7 Synchronization of Coupled Oscillators: Some Significant Results

After having discussed the general stability properties of synchronization states in complex networks, we review the most significant obtained results in the study of synchronization of different networked systems.

5.7.1 *Networks of phase oscillators*

As explained at the beginning of this chapter, there are several relevant situations, for instance in biology, where it is useful to consider the nodes of a given network as oscillatory systems. For this reason, the first network models were proposed with respect to limit cycle oscillators following the Kuramoto dynamics, which allows for both analytical and fast numerical simulations approaches.

Such works consider a generic graph where each node i $(i = 1, \cdots, N)$ is a planar rotor characterized by an angular phase, θ_i, and a natural or intrinsic frequency ω_i. Two oscillators interact if they are connected by an edge of the underlying graph. The individual dynamics of the ith node is described by:

$$\dot{\theta}_i = \omega_i + \sigma \sum_{j \in \mathcal{N}_i} \sin(\theta_j - \theta_i) \tag{5.14}$$

where $\mathcal{N}_i$ is the set of neighbors of i, and σ is the coupling strength, identical for all edges. The set of natural frequencies and the initial values of θ_i are in general randomly chosen from a given distribution [Kuramoto (1984); Strogatz (2000)].

The original Kuramoto model corresponds to the simplest case of globally coupled (complete graph) equally weighted oscillators, where the coupling strength is taken to be $\sigma = \varepsilon/N$ in order to warrant the smoothness of the model behavior also in the thermodynamic limit $N \to \infty$ [Kuramoto

(1984); Strogatz (2000)]. In this case, the onset of synchronization occurs at a critical value of the coupling strength $\varepsilon_c = 2/\pi g(\omega_0)$, where $g(\omega)$ is the distribution from which the natural frequencies are drawn, and ω_0 represents the mean frequency of the ensemble. The transition to synchronization is a second-order phase transition characterized by the order parameter:

$$r(t) = \left| \frac{1}{N} \sum_{j=1}^{N} e^{i\theta_j(t)} \right| . \tag{5.15}$$

When the two limits $N \to \infty$ and $t \to \infty$ are considered (and for $\varepsilon \geq \varepsilon_c$), the order parameter r behaves as $r \sim (\varepsilon - \varepsilon_c)^\beta$, with $\beta = 1/2$.

The mechanism through which synchronization emerges in the system is as follows. As the coupling takes very small values, the strength of the interactions is not enough to break the incoherence produced by the individual dynamics of each oscillator. When a critical value σ_c is overcome, some elements lock their relative phases and a cluster of synchronized nodes comes up. When the coupling is further increased, the population of oscillators is split into a partially synchronized state made up of oscillators locked in phase (that adds to r), and a group of nodes whose natural frequencies are too spread as to be part of the coherent pack. Eventually, after further increasing in σ, more and more nodes get entrained around the mean phase, and the system settles in a completely synchronous state (where $r \approx 1$). The globally coupled Kuramoto model has been extensively studied during the last several years, and a detailed and recent review on this subject can be found in [Acebrón (2005)].

With the growing interest in complex networks, several groups have moved the attention to investigate the synchronization phenomena of the Kuramoto model in complex wirings. [Moreno (2004a,b)] have studied numerically the conditions for the *onset* of synchronization in random scale-free networks. Specifically, they studied the Kuramoto model on top of SF networks and on top of small structures (motifs) that were relevant in different biological and social networks, with the aim of inspecting the critical point associated to the onset of synchronization, i.e., when small groups of synchronized oscillators first appear in the system.

The onset of synchronization (quantified by the order parameter given by Eq. (5.15)) for SF networks was reported to occur at a small, though non zero, value of the coupling strength, with a critical exponent around 0.5. In contrast with the all-to-all coupling configuration, for the complex topologies the critical point does not depend on the size of the system N.

Moreover, as it was discovered later on, the choice of the order parameter seems to be a crucial point when analyzing the conditions for the existence of the transition threshold.

Soon afterwards, several other authors [Lee (2004)] [Ichinomiya (2004); Restrepo (2004b)] have investigated the same problem from a theoretical perspective, as well as with numerical simulations.

The results do not fully clarify whether or not the critical point exists. The main difficulty comes from the fact that there is no a unique consensus about the set of differential equations describing the system dynamics, and what the order parameter should be. [Lee (2004)] reported the lack of a critical point in power law graphs when $2 < \gamma < 3$, which is recovered when the second moment of the distribution converges (that is for $\gamma > 3$). No estimates can be done for $\gamma = 3$ since there the relevant parameters of the system diverge.

The same qualitative behavior was reported in [Ichinomiya (2004)] [Restrepo (2004b)], where different analytical approaches were introduced.

The mean field theory [Lee (2004); Ichinomiya (2004)] predicts that the critical point is determined by the all-to-all Kuramoto value, σ_0, rescaled by the ratio between the first two moments of the degree distribution, $\sigma_{mf} = \sigma_0 \frac{\langle k \rangle}{\langle k^2 \rangle}$.

Recently, [Restrepo (2004b)] extended the results of [Lee (2004)] [Ichinomiya (2004)] by generalizing the mean-field approach.

Their results, supported by numerical simulations, show that the coupling strength at which the transition occurs is determined by the largest eigenvalue of the adjacency matrix. Moreover, they also point out that the mean-field approximation fails to predict the critical coupling strengths when the heterogeneity of the network increases (as, e.g., when γ gets closer or below to 3), because of the effect of the few nodes with a high degree (hubs) that are not properly taken into account in the mean field theory. At the same time, their time-averaged-theory seems to reproduce better the numerical estimates.

It is worth noticing that [Moreno (2004a,b)] have used the *classical* order parameter of Eq. (5.15), while the analysis of [Lee (2004); Ichinomiya (2004); Restrepo (2004b)] made use of a rescaled parameter, given by:

$$r(t) = \left| \frac{1}{\sum_l k_l} \sum_{j=1}^{N} k_l e^{i\theta_j(t)} \right| \tag{5.16}$$

This difference may be at the root of the apparent contradiction in the results for γ close to 3. In principle, there are reasons for the use of either one of the two parameters. Eq. (5.15) does not assign weights to the different connectivity classes, while Eq. (5.16) incorporates in the definition of the order parameter the connectivity of each node, which at first glance seems to be reasonable. However, this later assumption may have the effect of partially destroying the inherent heterogeneity of the underlying network by somehow homogenizing the contribution of each node to the synchronous state. Clearly, this issue needs to be addressed in more details in future works.

The other point of debate is whether or not one should rescale the coupling strength as $\sigma/\langle k\rangle$. While this is a constant factor for a given system size, it does not capture the main topological feature of scale-free networks: the fluctuations of the connectivity distribution. Thus one may ask whether dividing by $\langle k\rangle$ produces the same effects as normalizing by any other constant.

Finally, we briefly refers to clustering and modular synchronization of oscillator networks. It is believed that, as the route to complete synchronization is made up of groups of synchronized oscillators that grow and coalesce, one would be able to detect structural properties of the underlying network by just a fine tuning in the coupling strength. The effects of clustering have also been studied on different types of modular complex networks, finding that the synchronization transition crucially depends on the type of inter-modular connections [Oh (2005); McGraw (2005)]. The relaxation time $\langle\tau\rangle$ for synchronization of hubs was found to be shorter than that of less connected nodes [Moreno (2004a)]. In particular, for SF networks $\langle\tau\rangle \sim k^{-1}$. Hence, the more connected a node is, the more stable it is. Such power-law behavior points to an interesting result, namely, it is easier for an element with high k to get locked in phase with its neighbors than for a node linked to just a few others. Furthermore, the destabilization of a hub does not destroy the synchrony of the group it belongs to. On the contrary, the group formed by the hub's neighbors recruits it again. All these results indicate that it may be possible to use synchronization phenomena to unravel highly clustered structures embedded within the wiring of a complex networks.

5.7.2 *Networks of coupled oscillators*

The synchronizability of complex topologies has been largely studied in recent works [Nishikawa (2003); Barahona (2002); Wang (2002a)]

[Li (2003); Wu (2003); Lü (2004a,b); Lu (2004c); Hong (2004)]. Although some of these studies present slight differences in the assessment of the propensity for synchronization, all of them agree that the ability of a given network to synchronize is strongly ruled by the structure of connections.

In particular, some studies suggested that small world wirings always lead to enhance synchronization as compared with regular topologies [Lago-Fernández (2000); Gade (2000); Barahona (2002)] [Wang (2002a); Lü (2004a)]. Initially, this enhancement was attributed to the decreasing of the smaller average network distance between nodes. In fact, the statement that synchronization can be enhanced by the addition of *any* amount of random shortcuts is not always true [Pecora (1998a,b)]. The addition of arbitrary shortcuts to a locally regular coupled array may, under certain circumstances, result in a destabilization of the synchronized state [Jiang (2003)]. Moreover, synchronization can be affected by other quantities such as the degree distribution, characteristic path length, and betweenness centrality [Nishikawa (2003); Barahona (2002); Hong (2004)], and it was manifested that synchrony can be even deteriorated when increasing the heterogeneity in the connectivity distribution at the same average network distance [Nishikawa (2003)].

A recent numerical study of a SW network of Rössler oscillators revealed that phase synchronization may arise in the small world regime. Phase locking was found to depend on the coupling strength, the array size, or the distribution of the modes (natural frequencies) of each oscillator, as well as on the amount of random shortcuts in the connections [Li (2004)]. Although the Master Stability theory rigourously applies for coupled identical oscillators, numerical results confirm that the eigenratio also constitutes a good criterion to assess the propensity for synchronization of a network of nonidentical systems [Chavez (2005)].

The small world wiring seems therefore to yield an enhancement of synchronization (both in phase and amplitude), but nevertheless the relationship between the emergence of the SW regime (based on topological properties) and the arising of collective behaviors remains unclear. An example is the fact that the synchronization threshold does not coincide (neither in its value nor in its scaling properties) with that for the onset of the SW behavior [Barahona (2002)].

Synchronization of chaotic flows [Li (2003); Wang (2002b)] or phase synchronization of chaotic maps [Jalan (2003)] on top of scale-free networks was also demonstrated. Furthermore, networks and arrays of coupled maps were studied in the case in which the connections between units involve a time delay, to investigate the synchronization properties as a function of the underlying graph topology [Atay (2004)], or when the time delays among units are randomly selected [Masoller (2005)].

Complex wirings may display an intrinsical robustness again structural damage [Albert (2000)]. In scale free networks, for instance, the clustered organization of nodes as well as the connectivity patterns between such clusters, render the network structure tolerant to random removal of nodes but very vulnerable to targeted attacks [Albert (2000)]. This vulnerability not only affects static processes, but it is also directly related to the robustness of any collective dynamics emerging in such networks. Numerical simulations, for instance, have revealed that synchronization processes remain almost unchanged in scale free networks when a 5% of the total of nodes are randomly removed, while the network splits into isolated sub-networks (thus destroying synchronization) if only 1% of target (highly connected) nodes are removed [Lü (2004a); Wang (2002b)].

Finally, we briefly discuss the studies on the desynchronization mechanisms for networks. As far as coupled chaotic systems are considered, there exist many different scenarios by which synchronization is lost [Boccaletti (2002)]. For instance, a counterintuitive desynchronization scenario is characterized by a destabilization of the synchronized state produced by an increase of the coupling strength [Pecora (1998a); Heagy (1995)]. Known as short-wavelength bifurcation, this phenomenon is often observed in diffusively coupled arrays of chaotic oscillators, and is characterized by an excitation of the shortest spatial wavelength with an increasing of the coupling, which induces the desynchronization. A direct consequence of this phenomenon is the existence of an upper limit in the number of chaotic oscillators that can be synchronized, upon which the synchronized state of a given network becomes unstable [Heagy (1995); Pecora (1998b)].

In a recent related work, another desynchronization mechanism has also been studied in networks with arbitrary topology. Known as bubbling, this phenomenon arises when the synchronized state of non-identical systems is sporadically interrupted, yielding bursts of desynchronized motions [Ashwin (1994)]. The spatial patterns produced by these bursts were found to be ruled not only by the local dynamics of each node, but also by the structure of the coupling matrix [Restrepo (2004a)].

5.8 Conclusions

The present chapter presents only a snapshot of some relevant results of recent research on the interplay between network structure and dynamics. The research on synchronization phenomena in complex networks is far from being exhausted. On the contrary, the study of complex networks pervades in an interdisciplinary way all of the nonlinear science, and it has become one of the hottest topics, attracting the interest of a rapidly growing community of scientists. This scientific discipline will surely provide important insights into the mechanism that underlie real networks such as traffic networks, epidemiological models, or real brain networks.

References

Acebrón J. A., Bonilla L. L., Pérez Vicente C. J., Ritort F. and Spigler R. (2005). The Kuramoto model: a simple paradigm for synchronization phenomena. *Rev. Mod. Phys.* **77**, 137.

Afraimovich V.S., Verichev N.N. and Rabinovich M.I. (1986). Stochastic synchronization of oscillation in dissipative systems. *Radiophys. Quantum Electron.* **29**, 795.

Albert R., Jeong H., and Barabási A.-L. (2000). Attack and error tolerance of complex networks. *Nature* **406**, 378.

Albert R. and Barabási A.-L. (2002). Statistical mechanics of complex networks. *Rev. Mod. Phys.* **74**, 47.

Ashwin P., Buescu J. and Stewart I. (1994). Bubbling of attractors and synchronization of chaotic oscillators. *Phys. Lett.* **A193**, 126.

Atay F.M. , Jost J. and Wende A. (2004). Delays, connection topology, and synchronization of coupled chaotic maps. *Phys. Rev. Lett.* **92**, 144101.

Barabási A.-L., Albert R. and Jeong H. (1999). Mean-field theory for scale-free random networks. *Physica* **A272**, 173.

Barahona M. and Pecora L.M. (2002). Synchronization in Small-World Systems. *Phys. Rev. Lett.* **89**, 054101.

Barrat A., Barthélemy M., Pastor-Satorras R. and Vespignani A. (2004). The architecture of complex weighted networks. *Proc. Natl.Acad. Sci. U.S.A.* **101**, 3747 (2004).

Belykh V.N., Belykh I.V. and Hassler M. (2000). Hierarchy and stability of partially synchronous oscillations of diffusively coupled dynamical systems. *Phys. Rev.* **E62**, 6332.

Belykh, V.N., Belykh, I.V. and Hasler, M. (2004). Connection graph stability method for synchronized coupled chaotic systems. *Physica* **D195**, 159.

Belykh, I.V., Belykh, V.N. and Hasler, M. (2004). Blinking model and synchronization in small-world networks with a time-varying coupling. *Physica* **D195**, 188.

Belykh I., de Lange E. and Hasle, M. (2005). Synchronization of bursting neurons: what matters in the network topology. *Phys. Rev. Lett.* **94**, 188101.

Belykh, I., Hasler, M., Lauret, M. and Nijmeijer, H. (2005). Synchronization and graph topology. *Int J. Bifurcat. Chaos* **15**, 3423.

Berlow E.L. (1999). Strong effects of weak interactions in ecological communities. *Nature* **398**, 330.

Boccaletti S. and Valladares D.L. (2000). Characterization of intermittent lag synchronization. *Phys. Rev.* **E62**, 7497.

Boccaletti S., Kurths J., Valladares D.L., Osipov G. and Zhou C.S. (2002). The synchronization of chaotic systems. *Phys. Rep.* **366**, 1.

Boccaletti S., Latora V., Moreno Y., Chavez M. and Hwang D.-U. (2006). Complex Networks: Structure and Dynamics. *Phys. Rep.* **424**, 175.

Buzsáki G., Geisler C., Henze D.A. and Wang X.J.(2004). Circuit complexity and axon wiring economy of cortical interneurons. *Trends Neurosci.* **27**, 186.

Chavez M., Hwang D.-U., Amann A., Hentschel H.G. E. and Boccaletti S. (2005). Synchronization is enhanced in weighted complex networks. *Phys. Rev. Lett.* **94**, 218701.

Chen Y., Rangarajan G. and Ding M. (2003). General Stability Analysis of Synchronized Dynamics in Coupled Systems. *Phys. Rev.* **E67**, 026209.

Cvetkovic D. M. Doob M. and Sachs H. (1995). Spectra of Graphs: Theory and Applications, Johann Ambrosius Barth Verlag, Heidelberg.

Dorogovtsev S.N. and Mendes J.F.F. (2000). Evolution of reference networks with aging. *Phys. Rev.* **E62**, 1842.

Dorogovtesev S.N. and Mendes J.F.F. (2002). Evolution of networks. *Adv. Phys.* **51**, 1079.

Erdös P. and Rényi A. (1959). On Random Graphs I. *Publ. Math. Debrecen* **6**, 290.

Femat R. and Solis-Perales G. (1999). On the chaos synchronization phenomena. *Phys. Letts.* **A262**, 50.

Fink K.S., Johnson G., Carroll T.L. and Pecora L.M. (2000). Three coupled oscillators as a universal probe of synchronization stability in coupled oscillator arrays. *Phys. Rev.* **E61**, 5080.

Fujisaka H. and Yamada T. (1983). Stability Theory of Synchronized Motion in Coupled-Oscillator Systems . *Prog. Theor. Phys.* **69**, 32.

Gade P.M. and Hu C.-K. (2000). Synchronous chaos in coupled map lattices with small-world interactions. *Phys. Rev.* **E62**, 6409.

Gauthier D.J. and Bienfang J.C. (1996). Intermittent Loss of Synchronization in Coupled Chaotic Oscillators: Toward a New Criterion for High-Quality Synchronization. *Phys. Rev. Lett.* **77**, 1751.

Gerschgorin S.A. (1931). Über die Abgrenzung der Eigenwerte einer Matrix. *Izv. Akad. Nauk. USSSR, Otd. Fiz.-Mat. Nauk* **7**, 749.

Glass L. (2001). Synchronization and rhythmic processes in physiology *Nature* **410**, 277.

Goh K. I., Kahng B. and Kim D. (2001). Universal Behavior of Load Distribution in Scale-Free Networks. *Phys. Rev. Lett.* **87**, 278701.

Gong B., Yang L. and Yang K. (2005). Synchronization on Erdos-Renyi networks. *Phys. Rev.* **E72**, 037101.

Heagy J.F., Pecora L.M. and Carroll T.L. (1995). Short Wavelength Bifurcations and Size Instabilities in Coupled Oscillator Systems. *Phys. Rev. Lett.* **74**, 4185.

Hong H., Choi M.Y. and Kim B.J. (2002). Synchronization on small-world networks. *Phys. Rev.* **E65**, 026139.

Hong H., Kim B.J., Choi M.Y. and Park H. (2004). Factors that predict better synchronizability on complex networks. *Phys. Rev.* **E69**, 067105.

Hu G., Yang J. and Liu W. (1998). Instability and controllability of linearly coupled oscillators: Eigenvalue analysis. *Phys. Rev.* **E58**, 4440.

Hugenii C. (1673) *Horoloquium Oscilatorium*, Apud F. Muguet, Parisiis.

Hwang D.-U., Chavez M., Amann A. and Boccaletti S. (2005). Synchronization in complex networks with age ordering. *Phys. Rev. Lett.* **94**, 138701.

Ichinomiya T. (2004). Frequency synchronization in a random oscillator network. *Phys. Rev.* **E70**, 026116.

Jalan S. and Amritkar R.E. (2003). Self-Organized and Driven Phase Synchronization in Coupled Maps. *Phys. Rev. Lett.* **90**, 014101.

Jost J. and Joy M.P. (2002). Spectral properties and synchronization in coupled map lattices. *Phys. Rev.* **E65**, 016201.

Jiang Y., Lozada-Cassou M. and Vinet A. (2003). Synchronization and symmetry-breaking bifurcations in constructive networks of coupled chaotic oscillators. *Phys. Rev.* **E68**, 065201.

Kiss, I.Z., Zhai, Y. and Hudson, J.L. (2002). Emerging coherence in a population of chemical oscillators. *Science* **296**, 1676.

Klevecz R.R., Bolen J. and Durán O. (1992). Self-organization in biological tissues: analysis of asynchronous and synchronous periodicity, turbulence and synchronous chaos emergent in coupled chaotic arrays. *Int. J. Bifurcat. Chaos* **2**, 941.

Kocarev L. and Parlitz U. (1996). Generalized Synchronization, Predictability, and Equivalence of Unidirectionally Coupled Dynamical Systems. *Phys. Rev. Lett.* **76**, 1816.

Kocarev L. and Amato P. (2005). Synchronization in power-law networks. *Chaos* **15**, 02401.

Kuramoto Y. (1984). Chemical Oscillations, Waves, and Turbulence. Springer, Berlin.

Kuramoto Y. (1995). Scaling Behavior of Turbulent Oscillators with Non-Local Interaction. *Prog. Theor. Phys.* **94**, 321.

Kuramoto Y. and Nakao H. (1996). Origin of power-law spatial correlations in distributed oscillators and maps with nonlocal coupling. *Phys. Rev. Lett.* **76**, 4352.

Kuramoto, Y. and Nakao, H. (1997). Power-law spatial correlations and the onset of individual motions in self-oscillatory media with non-local coupling. *Physica* **D103**, 294.

Kwon O. and Moon H.-T. (2002). Coherence resonance in small-world networks of excitable cells. *Phys. Lett.* **A298**, 319.

Lago-Fernández L.F., Huerta R., Corbacho F. and Sigüenza J.A. (2000). Fast response and temporal coding on coherent oscillations in small-world networks. *Phys. Rev. Lett.* **84**, 2758.

Latora V. and Marchiori M. (2001). Efficient behavior of small-world networks. *Phys. Rev. Lett.* **87**, 198701.

Lee D.-S. (2004). Synchronization transition in scale-free networks: Clusters of synchrony. *Phys. Rev.* **E72**, 026208.

Li X. and Chen G. (2003). Synchronization and desynchronization of complex dynamical networks: An engineering viewpoint. *IEEE Trans. Circuits Syst. I* **50**, 1381.

Li C. and Chen G. (2004). Phase synchronization in small-world networks of chaotic oscillators . *Physica* **A341**, 73.

Lind P. G., Gallas J. A. C. and Herrmann H. J. (2004). Coherence in scale-free networks of chaotic maps. *Phys. Rev.* **E70**, 056207.

Liu Z., Lai Y.-C. and Hoppensteadt, F.C. (2001). Phase clustering and transition to phase synchronization in a large number of coupled nonlinear oscillators . *Phys. Rev.* **E63**, 055201.

Lü J.H., Yu X., Chen G.R. and Cheng D.Z. (2004). Characterizing the synchronizability of small-world dynamical networks. *IEEE Trans. Circuits Syst. I* **51**, 787.

Lü J.H., Yu X. and Chen G. (2004). Chaos synchronization of general complex dynamical networks. *Physica* **A334**, 281.

Lu W. and Chen T. (2004). Synchronization analysis of linearly coupled networks of discrete time systems. *Physica* **D198**, 148.

Manrubia S. C., Mikhailov A. S. and Zanette A. H. (2004). Emergence of Dynamical Order, World Scientific Lecture Notes in Complex Systems Vol. 2, World Scientific, Singapure, 348p.

Maródi M., d'Ovidio F. and Vicsek T.(2002). Synchronization of oscillators with long range interaction: Phase transition and anomalous finite size effects. *Phys. Rev.* **E66**, 011109.

Masoller C. and Martí A.C. (2005). Random Delays and the Synchronization of Chaotic Maps. *Phys. Rev. Lett.* **94**, 134102.

Massunaga M.S.O. and Bahiana M. (2002). Synchronization in large populations of limit cycle oscillators with long-range interactions . *Physica* **D168**, 136.

Matthews P.C., Mirollo R.E. and Strogatz S.H. (1991). Dynamics of a large system of coupled nonlinear oscillators. *Physica* **D52**, 293.

McCann K. A., Hastings A. and Huxel G. R. (1998). Weak trophic interactions and the balance of nature. *Nature***395**, 794.

McGraw P.N. and Menzinger M. (2005). Clustering and the synchronization of oscillator networks. *Phys. Rev.* **E72**, 015101R.

Moreno Y. and Pacheco A.F. (2004). Synchronization of Kuramoto Oscillators in Scale-Free Networks. *Europhys. Lett.* **68**, 603.

Moreno Y., Vázquez-Prada M. and Pacheco A.F. (2004). Fitness for Synchronization of Network Motifs. *Physica* **A343**, 279.

Motter A.E., Zhou C.S. and Kurths J. (2005). Enhancing complex-network synchronization. *Europhys. Lett.* **69**, 334.

Motter A.E., Zhou C. and Kurths J. (2005). Network synchronization, diffusion, and the paradox of heterogeneity. *Phys. Rev.* **E71**, 016116.

Newman M.E.J. (2003). The structure and function of complex networks. *SIAM Review* **45**, 167.

Niebur E., Schuster H.G., Kammen D.M. and Koch C. (1991). Oscillator-phase coupling for different two-dimensional network connectivities. *Phys. Rev.* **A44**, 6895.

Nishikawa T., Motter A.E., Lai Y.-C. and Hoppensteadt, F.C. (2003). Heterogeneity in oscillator networks: are smaller worlds easier to synchronize?. *Phys. Rev. Lett.* **91**, 014101.

Oh E., Rho K., Hong H. and Kahng B. (2005). Modular synchronization in complex networks. *Phys. Rev.* **E72**, 047101.

Osipov G.V., Pikovsky A.S., Rosemblum M.G. and Kurths J. (1997). Phase Synchronization Effects in a Lattice of Nonidentical Rössler Oscillators. *Phys. Rev.* **E55**, 2353.

Pecora L.M. and Carroll, T.L. (1990). Synchronization in chaotic systems. *Phys. Rev. Lett.* **64**, 821.

Pecora L.M. and Carroll T.L. (1998). Master Stability Functions for Synchronized Coupled Systems. *Phys. Rev. Lett.* **80**, 2109.

Pecora L.M. (1998). Synchronization conditions and desynchronizing patterns in coupled limit-cycle and chaotic systems. *Phys. Rev.* **E58**, 347.

Pikovsky A.S., Rosenblum M.G. and Kurths J. (1996) Synchronization in a Population of Globally Coupled Chaotic Oscillators. *Europhys. Lett.* **34**, 165.

Pikovsky, A., Rosemblum, M. and Kurths, J. (2001). Synchronization. A universal concept in nonlinear systems, Cambridge Nonlinear Science Series 12, Cambridge University Press, UK, 411p.

Pinto S.E. de S., Lopes S.R. and Viana R.L. (2002). Collective behavior in a chain of van der Pol oscillators with power-law coupling. *Physica* **A303**, 339.

Ramasco J.J., Dorogovtsev S. N. and Pastor-Satorras R. (2004). Self-organization of collaboration networks. *Phys. Rev.* **E70**, 036106.

Rangarajan G. and Ding M. (2002). Stability of Synchronized Chaos in Coupled Dynamical Systems. *Phys. Lett. A* **296**, 204.

Restrepo J.G., Ott E. and Hunt B.R. (2004). Spatial patterns of desynchronization bursts in networks. *Phys. Rev.* **E69**, 066215.

Restrepo J.G., Ott E. and Hunt B.R. (2004). The onset of synchronization in large networks of coupled oscillators. Preprint cond-mat/0411202.

Rogers J.L. and Wille L.T. (1996). Phase transitions in nonlinear oscillator chains. *Phys. Rev.* **E54**, R2193.

Rogister F., Thornburg Jr. K.S., Fabiny L. Möller M. and Roy R. (2004). Power-Law Spatial Correlations in Arrays of Locally Coupled Lasers. *Phys. Rev. Lett.* **92**, 093905.

Rosa E.R., Ott E. and Hess M.H. (1998). Transition to Phase Synchronization of Chaos. *Phys. Rev. Lett.* **80**, 1642.

Rosenblum M.G., Pikovsky A.S. and Kurths J. (1996). Phase synchronization of chaotic oscillators. *Phys. Rev.Lett.* **76**, 1804.

Rosenblum M.G., Pikovsky A.S. and Kurths J. (1997). From Phase to Lag Synchronization in Coupled Chaotic Oscillators. *Phys. Rev. Lett.* **78**, 4193.

Rulkov N.F., Sushchik M.M., Tsimring L.S. and Abarbanel H.D.I. (1995) Generalized synchronization of chaos in directionally coupled chaotic systems. *Phys. Rev.* **E51**, 980.

Rössler O.E. (1976). An equation for continuous chaos. *Phys. Lett.* **A57**, 397.

Sakaguchi H., Shinomoto S. and Kuramoto Y. (1987). Local and Grobal Self-Entrainments in Oscillator Lattices . *Prog. Theor. Phys.* **77**, 1005.

Sporns O., Chialvo D.R., Kaiser M. and Hilgetag C.C. (2004). Organization, development and function of complex brain networks. *Trends Cogn. Sci.* **8**, 418.

Stefański A., Wojewoda J., Kapitaniak T. and Yanchuk S. (2004). Simple estimation of synchronization threshold in ensembles of diffusively coupled chaotic systems. *Phys. Rev.* **E70**, 026217.

Strogatz S.H. and Mirollo R.E. (1988). Phase-locking and critical phenomena in lattices of coupled nonlinear oscillators with random intrinsic frequencies. *Physica* **D31**, 143.

Strogatz S. H., (2000). From Kuramoto to Crawford: exploring the onset of synchronization in populations of coupled oscillators. *Physica* **D143**, 1.

Strogatz S.H. (2001). Exploring complex networks. *Nature* **410**, 268.

Wang X.F. and Chen G. (2002). Synchronization in small-world dynamical networks. *Int. J. Bifurcat. Chaos* **12**, 187.

Wang X.F. and Chen G. (2002). Synchronization in scale-free dynamical networks: robustness and fragility. *IEEE Trans. Circuits Syst. I* **49**, 54.

Watts D.J. and Strogatz S. H. (1998). Collective dynamics of 'small-world' networks. *Nature* **393**, 440.

Winfree A. (1967). Biological rhythms and the behavior of populations of coupled oscillators. *J. Theor. Biol.* **16**, 15.

Wu C.W. and Chua L.O. (1995). Synchronization in an array of linearly coupled dynamical systems. *IEEE Trans. Circ. Syst.-I* **42**, 430.

Wu C.W. and Chua L.O. (1995). Application of graph theory to the synchronization in an array of coupled nonlinear oscillators. *IEEE Trans. Circ. Syst.-I* **42**, 494.

Wu C.W. and Chua L.O. (1996). On a conjecture regarding the synchronization in an array of linearly coupled dynamical systems. *IEEE Trans. Circ. Syst.-I* **43**, 161.

Wu C.W. (2003). Synchronization in Coupled Arrays of Chaotic Oscillators with Nonreciprocal Coupling. *IEEE Trans. Circuits Syst. I* **50**, 294.

Zaks M.A., Park E.-H., Rosenblum M.G. and Kurths J. (1999). Alternating Locking Ratios in Imperfect Phase Synchronization. *Phys. Rev. Lett.* **82**, 4228.

Zhan M., Hu G. and Yang J. (2000). Synchronization of chaos in coupled systems. *Phys. Rev.* **E62**, 2963.

Zhang Y., Hu G., Cerdeira H.A., Chen S., Braun T. and Yao Y. (2001). Partial synchronization and spontaneous spatial ordering in coupled chaotic systems. *Phys. Rev.* **E63**, 026211.

Zhao M., Zhou T., Wang B.-H. and Wang W.-X. (2005). Enhanced synchronizability by structural perturbations. *Phys. Rev.* **E72**, 057102.
Zheng Z., Hu G. and Hu B. (1998). Phase Slips and Phase Synchronization of Coupled Oscillators . *Phys. Rev. Lett.* **81**, 5318.
Zhou C.S. and Kurths J. (2002). Spatiotemporal coherence resonance of phase synchronization in weakly coupled chaotic systems. *Phys. Rev.* **E65**, 040101.
Zhou C., Motter A.E. and Kurths J. (2006). Universality in the synchronization of weighted random networks. *Phys. Rev. Lett.* **96**, 034101.

Chapter 6

Economic Sector Identification in a Set of Stocks Traded at the New York Stock Exchange: A Comparative Analysis

C. Coronnello, M. Tumminello, F. Lillo, S. Miccichè and R.N. Mantegna

We review some methods recently used in the literature to detect the existence of a certain degree of common behavior of stock returns belonging to the same economic sector. Specifically, we discuss methods based on random matrix theory and hierarchical clustering techniques . We apply these methods to a set of stocks traded at the New York Stock Exchange. The investigated time series are recorded at a daily time horizon.

All the considered methods are able to detect economic information and the presence of clusters characterized by the economic sector of stocks. However, different methodologies provide different information about the considered set. Our comparative analysis suggests that the application of just a single method could not be able to extract all the economic information present in the correlation coefficient matrix of a set of stocks.

6.1 Introduction

The correlation matrix is one of the major tools useful to study the collective behaviour of systems characterized by the presence of a large number of elements interacting with each other in a given period of time. Multivariate time series are detected and recorded both in experiments and in the monitoring of a wide number of physical, biological and economic systems.

The study of the properties of the correlation matrix has a direct relevance in the investigation of mesoscopic physical systems [Forrester (1994)], high energy physics [Demasure (2003)], information theory and communication [Moustakas (2000); Tworzydlo (2002); Skipetrov (2003)], investigation of microarray data in biological systems [Holter (2000)] [Alter (2000); Holter (2001)] and econophysics [Laloux (1999)] [Plerou (1999)] [Mantegna (1999); Maslov (2001); Pafka (2002)] [Malevergne (2004); Burda (1975)].

In the case of multivariate stock return time series, the correlation matrix may contain information about the economic sectors of the considered stocks [Mantegna (1999); Kullmann (2001); Bonanno (2000)] [Gopikrishnan (2001); Bonanno (2001); Kullmann (2002); Onnela (2002)] [Giada (2002); Bonanno (2003, 2004); Tumminello (2005)] [Coronnello (2005)].

Recent empirical and theoretical analysis have shown that this information can be detected by using a variety of methods. In this paper we review some of these methods based on Random Matrix Theory (RMT) [Gopikrishnan (2001)], correlation based clustering [Mantegna (1999)], and topological properties of correlation based graphs [Tumminello (2005)].

In this paper we perform a comparative analysis of the different techniques able to detect economic information out of the correlation matrix. In particular, our attention will be focused on the robustness of these methodologies with respect to statistical uncertainty and their capability in providing economic information. The review opens up with the discussion of the concepts of RMT. As explained below, this methodology can be used to select the eigenvalues and eigenvectors of the correlation matrix less affected by statistical uncertainty. Then we consider clustering techniques. In particular, we focus our attention on two correlation based hierarchical clustering procedures that can be used to obtain a reduced number of similarity measures representative of the whole original correlation matrix. These clustering procedures allow to obtain a number of similarity measures of the order of n, when starting from a similarity matrix with $n(n-1)/2$ distinct elements. The first clustering procedure we consider here is the single linkage clustering method that has been repeatedly used to detect a hierarchical organization of stocks. One advantage of using such correlation based clustering procedure is that it provides in a unique way a Minimum Spanning Tree (MST). The second clustering procedure is the average linkage which provides a different taxonomy and the last one is the PMFG. We also consider a recently introduced graph, the Planar Maximally Filtered Graph (PMFG) which extends the number of similarity measures associated to the graph with respect to the ones of the MST [Tumminello (2005)], although it has associated the same hierarchical tree.

The present investigation regards the transactions occured at the New York Stock Exchange (NYSE) in the year 2002 for a set of $n = 100$ highly capitalized stocks. The analysis is performed on the price returns computed at a daily time horizon.

The paper is organized as follows: in Section 2 we present the data set used in our empirical investigation. In Section 3 and Section 4 we illustrate the methods used to extract economic information from a correlation matrix of a stock set by using concepts and tools of RMT and hierarchical clustering. In Section 5 we illustrate a recently introduced technique which allows the construction of a graph, the Planar Maximally Filtered Graph (PMFG), obtained by imposing the topological constraint of planarity at each step of its construction. In Section 6 we draw our conclusions.

6.2 The Data Set

We investigate the statistical properties of price returns for $n = 100$ highly capitalized stocks traded at NYSE. In particular, we consider transactions occurred in year 2002. The empirical data are taken from the "Trades and Quotes" database, issued by the NYSE.

For each stock and for each trading day we consider the closure price. We compute daily returns as the difference of the logarithms of the closure prices of each successive trading day. To each of the 100 selected stocks an economic sector of activity is associated according to the classification scheme used in the web–site `http://finance.yahoo.com`. The economic sectors are reported in Table 6.1, together with the number of stocks belonging to each of them (third column).

The correlation coefficient between two stock return time series is defined as

$$\rho_{ij} = \frac{\langle r_i r_j \rangle - \langle r_i \rangle \langle r_j \rangle}{\sqrt{(\langle r_i^2 \rangle - \langle r_i \rangle^2)(\langle r_j^2 \rangle - \langle r_j \rangle^2)}} \qquad i, j = 1, \ldots, n \qquad (6.1)$$

where n is the number of stocks, i and j label the stocks, r_i is the logarithmic return defined by $r_i = \ln P_i(t) - \ln P_i(t - \Delta t)$, $P_i(t)$ is the value of the stock price i at the trading time t and Δt is the time horizon at which one computes the returns. In this work the correlation coefficient is computed between synchronous return time series. The correlation coefficient matrix is an $n \times n$ matrix whose elements are the correlation coefficients ρ_{ij}. An euclidean metric distance between pair of stocks can be rigorously

Table 6.1 Economic sectors of activity for 100 highly capitalized stocks traded at NYSE. The classification is done according to the methodology used in the web–site `http://finance.yahoo.com`. The second column contains the economic sector and the third column contains the number of stocks belonging to the sector.

	SECTOR	NUMBER
1	Technology	8
2	Financial	24
3	Energy	3
4	Consumer non-Cyclical	11
5	Consumer Cyclical	2
6	Healthcare	12
7	Basic Materials	6
8	Services	20
9	Utilities	2
10	Capital Goods	6
11	Transportation	2
12	Conglomerates	4

determined [Gower (1966)] by defining

$$d_{ij} = \sqrt{2(1 - \rho_{ij})} \tag{6.2}$$

With this choice d_{ij} fulfills the three axioms of a metric (i) $d_{ij} = 0$ if and only if $i = j$; (ii) $d_{ij} = d_{ji}$ and (iii) $d_{ij} \leq d_{ik} + d_{kj}$.

6.3 Random Matrix Theory

Random Matrix Theory [Metha (1990)] was originally developed in nuclear physics and then applied to many different fields. In the context of asset set management RMT is useful because it allows to quantify the statistical uncertainty, due to the finiteness of the time series under investigation, in the estimation of the correlation matrix. Let us consider n assets whose returns are described by n time series of length T. Let us also suppose that such returns are independent Gaussian random variables with zero mean and variance σ^2. In the limit $T \to \infty$, the correlation matrix of this set of variables is simply the identity matrix. However, when T is finite the correlation matrix will in general be different from the identity matrix. RMT allows to prove that for large values of T and n, with a fixed ratio $Q = T/n \geq 1$, the eigenvalue spectral density of the covariance matrix is given by

$$\rho(\lambda) = \frac{Q}{2\pi\sigma^2\lambda}\sqrt{(\lambda_{max}-\lambda)(\lambda-\lambda_{min})}, \tag{6.3}$$

where $\lambda_{min}^{max} = \sigma^2(1+1/Q \pm 2\sqrt{1/Q})$. The function $\rho(\lambda)$ is defined as the probability density function of eigenvalues. The above spectral density is different from zero in the interval $]\lambda_{min}, \lambda_{max}[$. In the case of a correlation matrix one can set $\sigma^2 = 1$. It is evident that the spectrum described by Eq. 6.3 is different from $\delta(\lambda - 1)$, i.e. from an identity correlation matrix. RMT is therefore a powerful tool able to quantify the role of the finiteness of the time series length on the spectral properties of the correlation matrix.

RMT has been applied to the investigation of correlation matrices of financial asset returns [Laloux (1999); Plerou (1999)] and it has been shown that the spectrum of a typical set can be divided in three classes of eigenvalues. The largest eigenvalue is totally incompatible with Eq. 6.3 and it is usually thought as describing the common behavior of the stocks composing the set. This fact leads to another working hypothesis according to which the part of correlation matrix which is orthogonal to the eigenvector corresponding to the first eigenvalue has a spectral density $\rho(\lambda)$ described by Eq. 6.3. Under these assumptions, the variance of the part not explained by the highest eigenvalue is given by $\sigma^2 = 1 - \lambda_1/n$ and such value is used to estimate of the variance to be used in Eq. 6.3 to compute $\tilde{\lambda}_{min}$ and $\tilde{\lambda}_{max}$. Following these lines, previous studies have shown that a fraction of the order of few percent of the eigenvalues are also incompatible with the RMT because they fall outside the interval $]\tilde{\lambda}_{min}, \tilde{\lambda}_{max}[$ computed by using the above estimate of σ^2. It is therefore natural to assume that these are the eigenvalues describing the economic information stored in the correlation matrix. The remaining large part of the eigenvalues is between $\tilde{\lambda}_{min}$ and $\tilde{\lambda}_{max}$ and thus one cannot say whether any information is contained in the corresponding eigenspace.

The fact that, under certain assumptions, RMT allows to identify the part of the correlation matrix containing economic information suggested some authors to use RMT for showing that the eigenvectors associated to eigenvalues not explained by RMT describe economic sectors. Specifically, the suggested method [Gopikrishnan (2001)] is the following. One computes the correlation matrix and finds the spectrum ranking the eigenvalues such that $\lambda_k > \lambda_{k+1}$. The eigenvector corresponding to λ_k is denoted $\mathbf{u}^k$. The set of investigated stocks is partitioned in S sectors $s = 1, 2, ..., S$ according to their economic activity (for example by using classification codes such as the one of the Standard Industrial Classification code or Forbes).

One then defines a $S \times n$ projection matrix $\mathbf{P}$ with elements $P_{si} = 1/n_s$ if stock i belongs to sector s and $P_{si} = 0$ otherwise. Here n_s is the number of stocks belonging to sector s. For each eigenvector $\mathbf{u}^k$ one computes

$$X_s^k \equiv \sum_{i=1}^{n} P_{si}[u_i^k]^2 \tag{6.4}$$

This number gives a measure of the role of a given sector s in explaining the composition of eigenvector $\mathbf{u}^k$. Thus when a given eigenvector has a large value of X_s^k for only one (or few) sector s, one can conclude that the eigenvector describes that economic sector (or a linear combination of a few of them). Finally, it is worth noting that the implementation of this method requires the *a priori* knowledge of the sector for each stock.

The above concepts and tools can be used to infer economic information about the set of stocks considered in Section 6.2. For a time horizon of one trading day the largest eigenvalue is $\lambda_1 = 39.2$ clearly incompatible with RMT and suggesting a driving factor common to all the stocks. This is usually interpreted to be the "market mode" as described in widespread market models, such as the Capital Asset Pricing Model. The analysis of the components of the corresponding eigenvector confirms this interpretation. In fact the mean component of the first eigenvector is 0.096 and the standard deviation is 0.027 showing that all the stocks contribute in a similar way to the eigenvector $\mathbf{u}^1$.

In our data $Q = T/n = 2.5$ and the threshold value λ_{max} without taking into account the first eigenvalue is $\lambda_{max} = 2.66$. This implies that RMT considers as signal only the first three eigenvalues λ_1, $\lambda_2 = 5.29$, and $\lambda_3 = 2.85$. On the other hand if we remove the contribution of the first eigenvalue with the procedure discussed in section 6.3 we get $\tilde{\lambda}_{max} = 1.62$, indicating that the first 9 eigenvalues could contain economic information. This result shows the importance of taking into account the role of the first eigenvalue.

Figure 6.1 shows X_s^k of Eq. 6.4 of the first 9 eigenvalues. Panel (a) shows that all the sectors contribute roughly in a similar way to the first eigenvector. On the other hand eigenvectors 2, 4, 5, and 9 are characterized by one prominent sector. Specifically, the second eigenvector shows a large contribution from the sector Consumer non-Cyclical ($s = 4$), the fourth eigenvector has a significant contribution from the Healthcare sector ($s = 6$), the fifth eigenvector is dominated by Utilities ($s = 9$) and Services ($s = 8$) and the ninth eigenvector has also a large peak for the sector Utilities ($s = 9$). The other eigenvectors do not show significant peaks, indicating

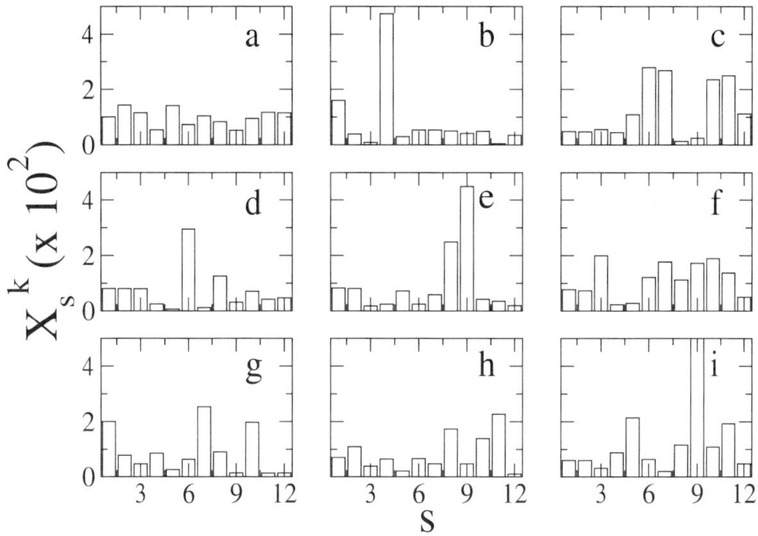

Fig. 6.1 X_s^k of Eq. 6.4 for the first nine eigenvectors of the correlation matrix of daily returns of 100 NYSE stocks. The order of sectors is the same as in Table 6.1.

that probably either these are eigenvectors mixing different groups or they are strongly affected by statistical uncertainty ("noise dressed").

6.4 Hierarchical Clustering Methods

Another approach used to detect the information associated to the correlation matrix is given by the correlation based hierarchical clustering analysis. Consider a set of n objects and suppose that a similarity measure, e.g. the correlation coefficient, between pairs of elements is defined. Similarity measures can be written in a $n \times n$ similarity matrix. The hierarchical clustering methods allow to hierarchically organize the elements in clusters. The result of the procedure is a rooted tree or dendrogram giving a quantitative description of the clusters thus obtained.

It is worth noting that hierarchical clustering methods can as well be applied to distance matrices.

A large number of hierarchical clustering procedures can be found in the literature. For a review about the classical techniques see for instance Ref. [Anderberg (1973)]. In this paper we focus our attention on the Single

Linkage Cluster Analysis (SLCA), which was introduced in finance in Ref. [Mantegna (1999)] and the Average Linkage Cluster Analysis (ALCA).

6.4.1 *Single linkage correlation based clustering*

The Single Linkage Cluster Analysis is a filtering procedure based on the estimation of the subdominant ultrametric distance [Rammal (1986)] associated with a metric distance obtained from the correlation coefficient matrix of a set of n stocks. This procedure, already used in other fields, allows to extract a hierarchical tree from a correlation coefficient matrix by means of a well defined algorithm known as nearest neighbor single linkage clustering algorithm [Mardia (1979)]. The methodology also allows to associate a MST to the hierarchical tree. Such association is essentially unique. By using this technique, it is therefore possible to reveal both topological (through the MST) and taxonomic (through the hierarchical tree) aspects of the correlation among stocks.

In the SLCA algorithm, at each step, when two elements or one element and a cluster or two clusters p and q merge in a wider single cluster t, the distance d_{tr} between the new cluster t and any cluster r is recursively determined as follows:

$$d_{tr} = \min\{d_{pr}, d_{qr}\} \tag{6.5}$$

thus indicating that the distance between any element of cluster t and any element of cluster r is the shortest distance between any two entities in clusters t and r. By applying iteratively this procedure $n-1$ of the $n(n-1)/2$ distinct elements of the original correlation coefficient matrix are selected.

The distance matrix obtained by applying the SLCA is an ultrametric matrix comprising $n-1$ distinct selected elements. The subdominant ultrametric is the ultrametric structure closest to the original metric structure [Rammal (1986)]. The ultrametric distance $d^<_{ij}$ between element i belonging to cluster t and element j belonging to cluster r is defined as the distance between clusters t and r. Ultrametric distances $d^<_{ij}$ are distances satisfying the inequality $d^<_{ij} \leq \max\{d^<_{jk}, d^<_{kj}\}$ stronger than the customary triangular inequality $d_{ij} \leq d_{ik} + d_{kj}$ [Rammal (1986)]. The SLCA has associated a correlation matrix which is associated to the subdominant ultrametric distance matrix obtained from the original correlation coefficient matrix. It can be obtained starting from the ultrametric distances $d^<_{ij}$ and making use of Eq. 6.2.

The MST is a graph without loops connecting all the n nodes with the shortest $n-1$ links amongst all the links between the nodes. The selection of these $n-1$ links is done according to some widespread algorithm [Papadimitriou (1982)] and can be summarized as follows:

(1) Construct an ordered list of pairs of stocks L_{ord}, by ranking all the possible pairs according to their distance d_{ij}. The first pair of L_{ord} has the shortest distance.
(2) The first pair of L_{ord} gives the first two elements of the MST and the link between them.
(3) The construction of the MST continues by analyzing the list L_{ord}. At each successive stage, a pair of elements is selected from L_{ord} and the corresponding link is added to the MST only if no loops are generated in the graph after the link insertion.

Different elements of the list are therefore iteratively included in the MST starting from the first two elements of L_{ord}. As a result, one obtains a graph with n vertices and $n-1$ links. For a didactic description of the method used to obtain the MST one can consult Ref. [Mantegna (2000)]

In Ref. [Gower (1969)] the procedure briefly sketched above has been shown to provide a MST which is associated to the same hierarchical tree of the SLCA. In Ref. [Tumminello (2006)] it is proved that the correlation matrix obtained by the SLCA is always positive definite when all the elements of the obtained correlation matrix are non negative. This condition is rather common in financial data.

The effectiveness of the SLCA in pointing out the hierarchical structure of the investigated set has been shown by several studies [Mantegna (1999); Bonanno (2000, 2001); Kullmann (2002)] [Onnela (2002); Bonanno (2003, 2004); Micciché (2003)] [Di Matteo (2004)].

The methodology sketched above can again be applied to the set of stocks considered in Section 6.2. The results obtained by using the SLCA for the daily returns are summarized in Fig. 6.2 and Fig. 6.3 that show the hierarchical tree and the MST, respectively.

The hierarchical tree shows that there exists a significant level of correlation in the market, and also in many cases clustering can be observed. For example, the 60_{th} and 61_{st} stocks of Fig. 6.2, *Freddie Mac* (FRE) and *Fannie Mae* (FNM), belonging to the Financial sector, are linked together at an ultrametric distance $d^{<} \approx 0.45$ corresponding to a correlation coefficient as high as $\rho = 0.90$. We focus here our attention on the two sectors with the largest number of stocks, which are the Financial sector ($s = 2$)

and the Services sector ($s = 8$). Panel (a) of Fig. 6.2, where the stocks of the Financial sector are highlighted, gives an example in which most of the stocks belonging to the same economic sector are clustered together. In fact, a cluster including 18 stocks from position 1 to position 18 can be observed. The remaining six stocks of the Financial sector are *All State Corp* (ALL) at position 29, *Aflac* (AFL) at position 41, FRE at position 60, FNM at position 61, *Well Point* (WLP) at position 94 and *CIGNA* (CI) at position 98. ALL and AFL are insurance companies, FNM and FRE are mortgage companies while WLP and CI provide health or commercial benefits for the employees of other companies. On the contrary, all the 18 stocks that cluster together are relative to banks, financial institutions or companies that provide financial services, with the only exception of *Progressive Corp* (PGR) at position 16 which is an insurance company. This result, obtained for the Financial sector, is representative of the capability of SLCA to evidentiate economic subsectors. In panel (b) of Fig. 6.2 the stocks of the Services sector are highlighted. The Services sector gives an example of the case in which stocks belonging to the same economic sector are poorly clustered. In fact, the biggest cluster is composed of only four stocks. Specifically, the cluster is formed by *Lowes Companies* (LOW), *Home Depot* (HD), *Kohls Corporation* (KSS) and *Wal-Mart Stores* (WMT) located from position 55 to position 58 in the figure. It is worth noting that all of them are classified as retail companies at the level of economic subsector. This results suggests that economic subsectors are highly characterizing for Services companies. Another economic sector that shows a significant level of clustering is the Basic Materials sector. The smallest economic sectors, i.e. Energy, Consumer Cyclical and Utilities are completely clustered. The stocks of the Technology, Consumer Non Cyclical and Healthcare economic sectors are not fully clustered. It is worth noting that the results regarding the Consumer Non Cyclical and the Healthcare economic sectors are different from those obtained by the RMT.

The MST partly confirms the above results. In Fig. 6.3 the stocks belonging to the Financial economic sector (black circles) and the Services economic sector (gray circles) are indicated and the results agree with those shown by the hierarchical tree. However, the MST gives slightly different informations from the hierarchical tree when considering the Technology, the Consumer Non Cyclical and the Healthcare economic sectors. In order to illustrate this point, in Fig. 6.4 the stocks belonging to the Technology economic sector (black circles) and the Consumer Non Cyclical economic sector (gray circles) are indicated. The figure shows that 6 out of 8 stocks of

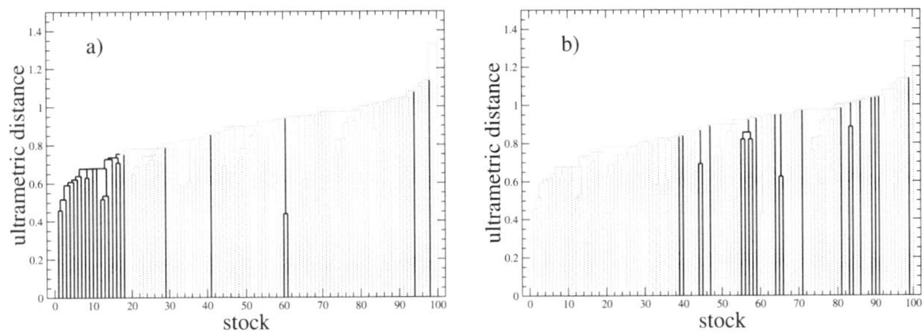

Fig. 6.2 Hierarchical tree obtained by using the SLCA starting from the daily price returns of 100 highly capitalized stocks traded at NYSE. Only transactions occurred in year 2002 are considered. In panel (a) the Financial economic sector is highlighted. In panel (b) the Services economic sector is highlighted.

the Technology sector are gathered around *Analog Devices Inc* (ADI), while 8 out of 11 stocks of the Consumer Non Cyclical sector are gathered around *Procter & Gamble* (PG). This observation suggests that the MST is carrying information which is additional to the one present in the hierarchical tree.

Finally, the topology of the MST shows the existence of several stocks with relatively large degree (number of links with other elements). This is incompatible with the hypothesis that the set of stocks is described by a single factor model [Bonanno (2003)].

6.4.2 *Average linkage correlation based clustering*

The Average Linkage Cluster Analysis is another hierarchical clustering procedure [Anderberg (1973)] that can be described by considering either a similarity or a distance measure. As in the previous case, here we consider the distance matrix $\mathbf{D}$. The following procedure performs the ALCA giving as an output a rooted tree and an ultrametric matrix $\mathbf{D}^<$ of elements $d^<_{ij}$:

(1) Set $\mathbf{T}$ as the matrix of elements such that $\mathbf{T} = \mathbf{D}$.
(2) Select the minimum distance t_{hk} in the distance matrix $\mathbf{T}$. Note that after the first step of construction h and k can be simple elements (i.e. clusters of one element each) or clusters (sets of elements).
(3) Merge cluster h and cluster k into a single cluster, say h. The merging operation identifies a node in the rooted tree connecting clusters h and k at the distance t_{hk}. Furthermore to obtain the ultrametric matrix it is sufficient that $\forall\, i \in h$ and $\forall\, j \in k$ one sets $d^<_{ij} = d^<_{ji} = t_{hk}$.

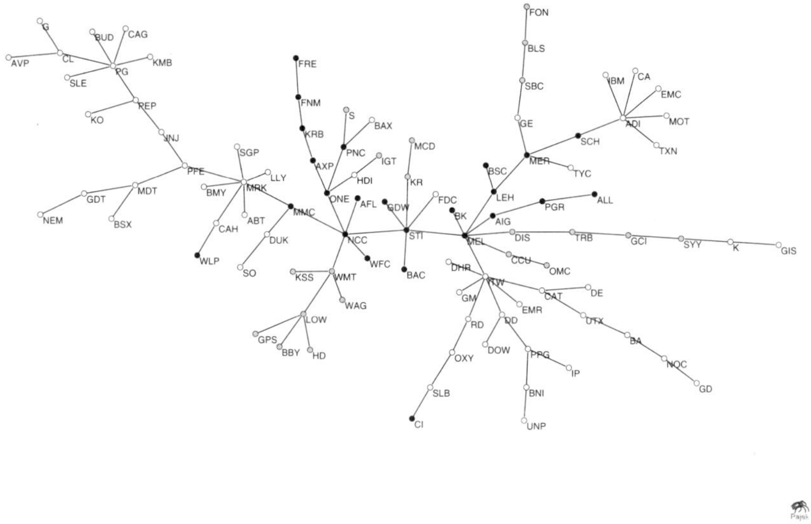

Fig. 6.3 MST obtained starting from the daily price returns of 100 highly capitalized stocks traded at NYSE. Only transactions occurred in year 2002 are considered. The Financial economic sector (black) and the Services (gray) economic sector are highlighted.

(4) Redefine the matrix $\mathbf{T}$:

$$\begin{cases} t_{hj} = \dfrac{N_h\, t_{hj} + N_k\, t_{kj}}{N_h + N_k} & \text{if } j \neq h \,\text{and}\, j \neq k \\ \\ t_{ij} = t_{ij} & \text{otherwise,} \end{cases}$$

where N_h and N_k are the number of elements belonging respectively to the cluster h and to the cluster k. Note that if the dimension of $\mathbf{T}$ is $m \times m$ then the dimension of the redefined $\mathbf{T}$ is $(m-1) \times (m-1)$ because of the merging of clusters h and k into the cluster h.

(5) If the dimension of $\mathbf{T}$ is bigger than one then go to step 2 else Stop.

By replacing point 4 of the above algorithm with the following item

4. Redefine the matrix $\mathbf{T}$:

$$\begin{cases} t_{hj} = Min\,[t_{hj}, t_{kj}] & \text{if } j \neq h \,\text{and}\, j \neq k \\ t_{ij} = t_{ij} & \text{otherwise,} \end{cases}$$

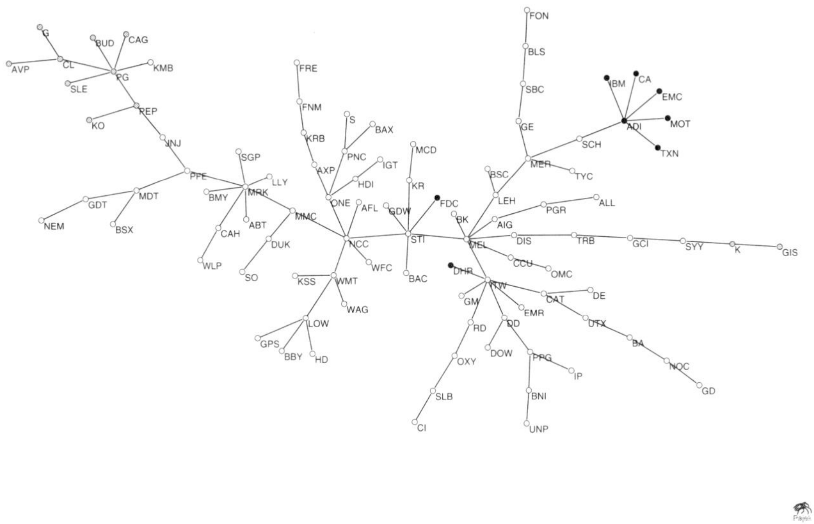

Fig. 6.4 MST obtained starting from the daily price returns of 100 highly capitalized stocks traded at NYSE. Only transactions occurred in year 2002 are considered. The Technology economic sector (black) and the Consumer Non Cyclical (gray) economic sector are highlighted.

one obtains an algorithm performing the SLCA which is therefore equivalent to the one described in the previous section. The algorithm can be easily adapted for working with similarities instead of distances. It is just enough to exchange the distance matrix $\mathbf{D}$ with a similarity matrix (for instance the correlation matrix) and replace the search for the minimum distance in the matrix $\mathbf{T}$ in point 2 of the above algorithm with the search for the maximal similarity.

It is worth noting that the ALCA can produce different hierarchical trees depending on the use of a similarity matrix or a distance matrix. More precisely, different dendrograms can result for the ALCA due to the non linearity of the transformation of Eq. 6.2. This problem does not arise in the SLCA because Eq. 6.2 is a monotonic transformation and then ranking the connections with respect to correlations from the greatest to the smallest gives the same results than ranking the connections with respect to distances from the smallest to the greatest.

Unlike the case of the SLCA, it is not possible to uniquely associate a spanning tree to the hierarchical tree obtained by using the ALCA. This is due to the fact that in the ALCA the distance between clusters is defined as the mean distance between the elements of the clusters. Such choice hides the possibility of defining in a unique way what is the path connecting two stocks.

As much as in the case of SLCA we can apply the above methodology to the empirical data described in Section 6.2. Here we analyze the dendrogram of Fig. 6.5 obtained by applying the ALCA to the correlation based distance matrix of the daily returns. Once again, to provide representative examples we focus our attention to the two sectors with the largest number of stocks.

As in Fig. 6.2, in panel (a) of Fig. 6.5 the black lines are identifying stocks of the Financial sector. The figure shows that 16 out of 24 stocks belonging to the Financial sector cluster together. The distance between the elements of this cluster are lower than the average distance. Exceptions are *Golden West Financial Corp* (GDW) at position 18, AFL at position 20, ALL at position 39, *Charles Schwab Corp* (SCH) at position 46, FRE at position 55, FNM at position 56, WLP at positions 96 and CI and at positions 97. Interestingly, AFL, ALL, FRE, FNM, WLP and CI are distant from the observed cluster also when considering the SLCA, as shown in panel (a) of Fig. 6.2. In panel (b) of Fig. 6.5 the black lines are identifying the 20 stocks belonging to the Services sector. Here a cluster of five stocks, from position 50 to position 54 is observed. Moreover, stocks at positions from 50 to 53 are the same ones that can be found in the cluster observed

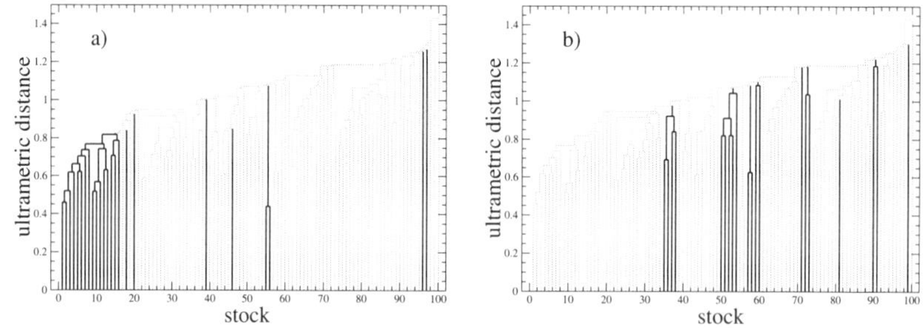

Fig. 6.5 Dendrogram associated to the ALCA performed on daily returns of a set of 100 stocks traded at NYSE in 2002. Panel (a): The black lines are identifying stocks belonging to the Financial sector. Panel (b): The black lines are identifying stocks belonging to the Services sector.

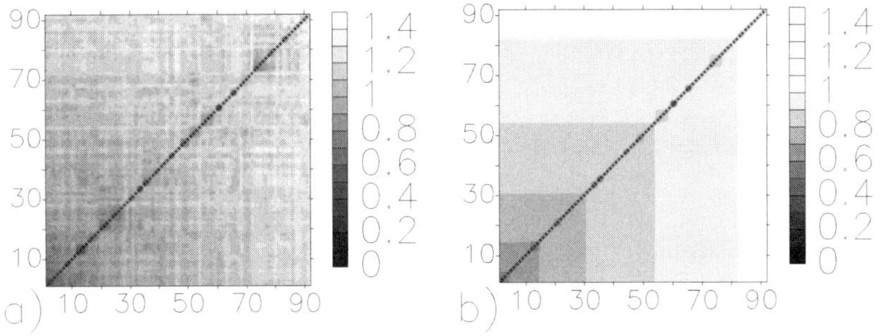

Fig. 6.6 Contour plots of the original distance matrix (panel (a)) and of the one associated to the ultrametric distance (panel (b)) obtained by using the SLCA for the daily price returns of 100 highly capitalized stocks traded at NYSE. Only transactions occurred in year 2002 are considered. Here the stocks are identified by a numerical label ranging from 1 to 100 and ordered according to the hierarchical tree of Fig. 6.2. The figure gives a pictorial representation of the amount of information which is filtered out by using the SLCA.

in panel (b) of Fig. 6.2, from position 55 to position 58. The fifth stock in the cluster is *WalGreen Company* (WAG) which is again a company classified as retail at the level of economic subsector. Another cluster of four stocks is observed from position 35 to position 38. Interestingly, the two stocks at positions 35 and 36 belong to the Printing & Publishing economic subsector, while the two companies at positions 37 and 38 belong to the Broadcasting economic subsector. This result provides a different example which is representative of the capability of hierarchical clustering methods to evidentiate economic subsectors.

A comparison of the results obtained by using the SLCA and the ALCA shows a substantial agreement between the output of these two methods. However, a refined comparison shows that the ALCA provides a more structured hierarchical tree. In Fig. 6.6 and Fig. 6.7 we show a graphical representation of the original distance matrix done in terms of a contour plot. In the contour plot the gray scale represents the values of distances among stocks. In the figures we use as stock order the one obtained by SLCA and ALCA respectively. In both cases we also show the associated ultrametric matrices. A direct comparison of the ultrametric distance matrices confirms that ALCA is more structured than SLCA. Conversely, the SLCA selects elements of the matrix with correlation values greater than the ones selected by ALCA and then less affected by statistical uncertainty.

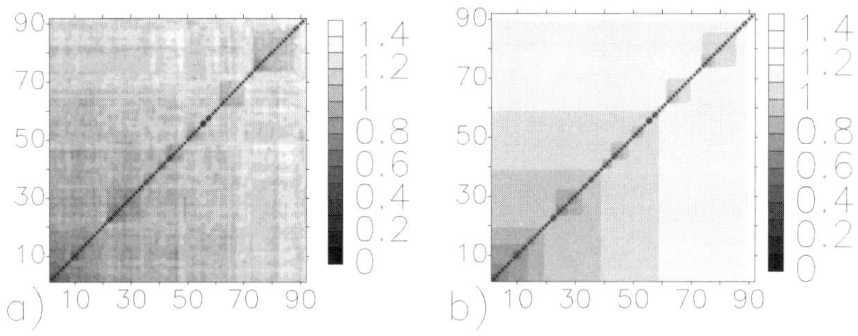

Fig. 6.7 Contour plots of the original distance matrix (panel (a)) and of the one associated to the ultrametric distance (panel (b)) obtained by using the ALCA for the daily price returns of 100 highly capitalized stocks traded at NYSE. Only transactions occurred in year 2002 are considered. Here the stocks are identified by a numerical label ranging from 1 to 100 and ordered according to the hierarchical tree of Fig. 6.5. The figure gives a pictorial representation of the amount of information which is filtered out by using the ALCA.

6.5 The Planar Maximally Filtered Graph

The Planar Maximally Filtered Graph has been introduced in a recent paper [Tumminello (2005)]. The basic idea is to obtain a graph retaining the same hierarchical properties of the MST, i.e. the same hierarchical tree of SLCA, but allowing a greater number of links and more complex topological structures than the MST. The construction of the PMFG is done by relaxing the topological constraint of the MST construction protocol of section 6.4.1 according to which no loops are allowed in a tree. Specifically, in the PMFG a link can be included in the graph if and only if the graph with the new link included is still planar. A graph is planar if and only if it can be drawn on a plane without edge crossings [West (2001)].

The first difference between MST and PMFG is about the number of links, which is $n-1$ in the MST and $3(n-2)$ in the PMFG. Furthermore loops and cliques are allowed in the PMFG. A clique of r elements, r-cliques, is a subgraph of r elements where each element is linked to each other. Because of the Kuratowski's theorem [West (2001)] only 3-cliques and 4-cliques are allowed in the PMFG. The study of 3-cliques and 4-cliques is relevant for understanding the strength of clusters in the system [Tumminello (2005)] as we will see below in the empirical applications. This study can be done through a measure of the intra-cluster connection strength [Tumminello (2005)]. This measure is obtained by considering a specific sector composed by n_s elements and indicating with c_4 and c_3 the number of 4-cliques and 3-cliques exclusively composed by elements of the sector. The connection strength q_s of the sector s is therefore defined as

$$q_s^4 = \frac{c_4}{n_s - 3}, \qquad q_s^3 = \frac{c_3}{3n_s - 8}, \tag{6.6}$$

where we distinguish between the connection strength evaluated according to 4-cliques q_s^4 and 3-cliques q_s^3. The quantities $n_s - 3$ and $3n_s - 8$ are normalizing factors. For large and strongly connected sectors both the measures give almost the same result [Tumminello (2005)]. When small sectors are considered the quantity q_s^3 is more significant than q_s^4. Consider for instance a sector of 4 stocks. In this case q_s^4 can assume the value 0 or 1, whereas q_s^3 can assume one the 5 values 0, 0.25, 0.5, 0.75 and 1, giving a measure of the clustering strength less affected by the quantization error. Note that in this case if q_s^3 assumes one of the values 0, 0.25, 0.5 and 0.75 then q_s^4 is always zero. Concerning the hierarchical structure associated to the PMFG it has been shown in Ref. [Tumminello (2005)] that at any step in the construction of the MST and PMFG, if two elements are connected via at least one path in one of the considered graphs, then they also are connected in the other one. This statement implies that i) the MST is always contained in the PMFG and ii) the hierarchical tree associated to both the MST and PMFG is the one obtained from the SLCA.

In summary the PMFG is a graph retaining more information about the system than the MST, the information being stored in the included new links and in the new topological structures allowed. i.e. loops and cliques.

The capability of the PMFG of exploiting economic information from a given correlation matrix can be illustrated by considering the set of stocks of Section 6.2. Here we analyze the topological properties of the PMFG of Fig. 6.8 obtained from the distance matrix of daily returns of the stock set.

In the figure we again point out the behavior of stocks belonging to the Financial and Services sectors. From the figure we can observe that the Financial sector (black circles) is strongly intra-connected (black thicker edges) whereas for the sector of Services (gray circles) we find just a few intra-sector connections (gray thicker edges). These results agree with the ones observed with the SLCA and the ALCA. The advantage of the study of the PMFG is that, through it, we can perform a quantitative analysis

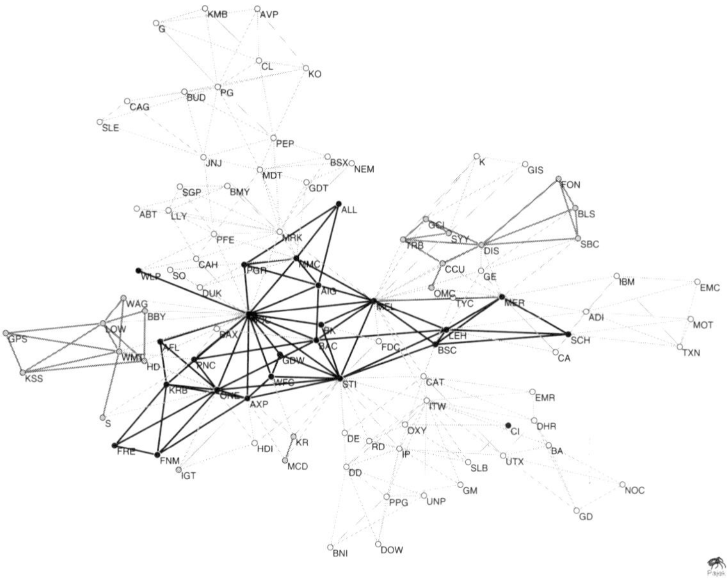

Fig. 6.8 PMFG obtained from daily returns of a set of 100 stocks traded at NYSE in 2002. Black circles are identifying stocks belonging to the Financial sector. Gray circles are identifying stocks belonging to the Services sector. Other stocks are indicated by empty circles. Black thicker lines are connecting stocks belonging to the Financial sector. Gray thicker lines are connecting stocks belonging to the Services sector.

of this behavior, by using the intra–cluster connection strength defined in section 6.5.

In Table 6.2 the connection strength is evaluated for all the sectors present in the set. The Financial sector has $q_2^4 \approx 0.91$ and $q_2^3 \approx 0.91$. This last value is the second biggest value, after the one associated to the Energy. The Energy sector is composed by 3 stocks which are all connected within them so that $q_3^3 = 1$. The results for the Financial and the Energy sectors agree with the ones shown by both hierarchical trees and the MST. For the Healthcare sector we find $q_6^3 \cong 0.68$, thus showing again an high degree of intra-cluster connection. A significant degree of intra-cluster connection can also be observed for the Consumer Non Cyclical sector ($q_3^4 \cong 0.60$). Both results were also shown by the MST and the hierarchical tree obtained by applying the ALCA. Part of this cluster is also detected by SLCA.

Among the stocks of the Financial sector, *National City Corp* (NCC), *Sun Trust Banks Inc* (STI) and *Mellon Financial Corp* (MEL) are charac-

Table 6.2 Intra-sector connection strength (daily returns)

SECTOR	n_s	$q_s^4 = c_4/[n_s - 3]$	$q_s^3 = c_3/[3\,n_s - 8]$
Technology	8	$1/5 = 0.20$	$5/16 = 0.31$
Financial	24	$19/21 \approx 0.91$	$58/64 \approx 0.91$
Energy	3	$--$	$1/1 = 1.00$
Consumer non-Cyclical	11	$4/8 = 0.50$	$15/25 \approx 0.60$
Consumer Cyclical	2	$--$	$--$
Healthcare	12	$5/9 \approx 0.56$	$19/28 \approx 0.68$
Basic Materials	6	$1/3 \approx 0.33$	$4/10 = 0.40$
Services	20	$3/17 \approx 0.18$	$16/52 \approx 0.31$
Utilities	2	$--$	$--$
Capital Goods	6	$0/3 = 0.00$	$1/10 = 0.10$
Transportation	2	$--$	$--$
Conglomerates	4	$0/1 = 0.00$	$0/4 = 0.00$

terized by high values of their degree. This fact implies that the Financial sector is strongly connected both within the sector and with other sectors. In particular, NCC is the center of the biggest star in the graph with a degree of 26 (15 are links to stocks of the same sector). On the contrary, the sector of Services is poorly intra-connected: $q_8^4 = 0.18$ and $q_8^3 \cong 0.31$. This sector is also poorly connected to other sectors. In fact, the highest degree of a stock belonging to this sector is 9 and it is observed for WMT and *Walt Disney* (DIS). The topology of the PMFG shows that the stocks of the Services sector are mainly gathered in two different subgraphs, clearly visible on the left and the right regions of the graph. The subgraph containing WMT is composed by retail companies. The other one is composed by stocks belonging to subsectors which are different although related the one to each other. In fact, *Sprint Corp FON Group* (FON), *Bellsouth Corp.* (BLS) and *SBC Communications* (SBC) belong to the Communication Services subsector, DIS and *Clear Channel Communications* (CCU) belong to the Broadcasting subsector, *Tribune Company* (TRB) and *Gannett Co* (GCI) belong to the Printing & Publishing subsector and *Omnicom Group* (OMC) belongs to the Advertising subsector. The only exception seems to be *Sysco Corp* (SYY) which belongs to the Retail Grocery subsector. This result provides an example representative of the way in which the PMFG recognizes economic subsectors. The fact that the connection strenghts q_8^4 and q_8^3 assume low values is indeed explained by the fact that the PMFG detects two subclusters of the same economic sector which are not linked to each other. This is in agreement with the results obtained by using the ALCA and SLCA. In fact, in the case of ALCA we observed two distinct clusters, one of retail companies and another one containing Printing & Publishing and Broadcasting companies. In the case of the SLCA, which provides a less structured hierarchical tree, only one cluster of retails companies was observed.

In conclusion, we observe that the Financial sector is strongly intra-connected and strongly connected with other sectors. On the other hand, the sector of Services is poorly intra-connected and poorly interacting with other sectors.

6.6 Conclusions

We observe that all methodologies considered show that the investigated system is structured at daily time horizons. All the methods considered in the present paper are able to detect information about economic sectors of stocks starting from the synchronous correlation coefficient matrix of return time series, although the degree of efficiency in the detection can be different for different techniques.

In fact, our comparative study shows that the considered methods provide different information about the system. For example, at one day time horizon the method based on RMT predominantly associates the eigenvectors of the highest eigenvalues which are not affected by statistical uncertainty respectively to the market factor (first eigenvalue and eigenvector), the Consumer non-Cyclical sector (second eigenvalue), the Healthcare sector (fourth eigenvalue), the Utilities sector (fifth and nineth eigenvalue) and the Services sector (fifth eigenvalue). In the present case, RMT does not provide information about the existence and strength of economic relation between stocks belonging to the sectors of Technology, Financial, Energy, Consumer Cyclical, Basic Materials, Capital Goods, Transportation and Conglomerates economic sectors.

A detailed investigation of the hierarchical trees obtained by the SLCA and ALCA shows that these methods are able to efficiently detect various clusters. They can be different from those obtained with the RMT. For example, the Financial sector is well detected by the SLCA and ALCA, while the Technology, the Consumer Non-Cyclical and the Healthcare economic sectors are not well detected at the level of the hierarchical tree. However, it must be mentioned that in the case of the SLCA, the MST can provide distinct yet not independent information about the level of clustering. In particular, the Technology, the Consumer Non-Cyclical and the Healthcare economic sectors are well detected in the MST. The information provided by the MST is usually confirmed by the quantitative analysis of the PMFG performed by computing the intrasector connection strength.

Our comparative analysis of the hierarchical clustering methods also shows that SLCA is providing information about the highest level of correlation of the correlation matrix whereas the ALCA averages this information within each considered cluster. In this way the average linkage clustering is able to provide a more structured information about the hierarchical organization of the stocks of a set, although it is more affected by statistical uncertainty.

In summary, we believe that our empirical comparison of different methods provides an evidence that RMT and hierarchical clustering methods are able to point out information present in the correlation matrix of the investigated system. The information that is detected with these methods is in part overlapping but in part specific to the selected investigating method. In short, all the approaches detect information but not exactly the same one. For this reason an approach that simultaneously makes use of several of these methods may provide a better characterization of the investigated system than an approach based on just one of them.

References

Forrester P.J. and Hughes T.D., J. Math. Phys. **35**, 6736 (1994).

Demasure Y. and Janik R.A., Phys. Lett. B **553**, 105 (2003).

Moustakas A.L. et al, Science **287**, 287 (2000).

Tworzydlo J. and BeenakkerC.W.J., Phys. Rev. Lett. **89**, 043902 (2002).

Skipetrov S.E., Phys. Rev. E **67**, 03362 (2003).

Holter N.S. *et al.*, Proc. Nat. Acad. Sci. USA **97**, 8409 (2000).

Alter O., Brown P.O. and Botstein, Proc. Nat. Acad. Sci. USA **97**, 10101 (2000).

Holter N.S. et al, Proc. Nat. Acad. Sci. USA **98**, 1693 (2001).

Laloux L., Cizeau P.,Bouchaud J.P., and Potters M., Phys. Rev. Lett. **83**, 1468 (1999).

Plerou V., Gopikrishnan P. ,Rosenow B., Amaral L.A.N. and Stanley H.E., Phys. Rev. Lett. **83**, 1471 (1999).

Mantegna R.N., Eur. Phys. J. B **11**, 193 (1999).

Maslov S. and Zhang Y.-C., Phys. Rev. Lett. **87**, 248701 (2001).

Pafka S. and Kondor I., Eur. Phys. J. B **27**, 277 (2002).

Malevergne Y. and Sornette D., Physica A **331**, 660 (2004).

Burda Z. et al, Physica A **343**, 295 (2004).

Kullmann L., Kertesz J. and Mantegna R.N., *Physica A* **287**, 412 (2001).

Bonanno G., Vandewalle N. andMantegna R.N., Phys. Rev. E **62**, R7615 (2000).

Gopikrishnan P., Rosenow B., Plerou V. and Stanley H.E., Phys. Rev. E **64**, 035106 (2001).
Bonanno G. , Lillo F. and Mantegna R.N., *Quantitative Finance* **1**, 96 (2001).
Kullmann L., Kertesz J. and Kaski K., Phys. Rev. E **66**, 026125 (2002).
Onnela J.P., Chakraborti A., Kaski K. and Kertesz J., *Eur. Phys. J. B* **30**, 285 (2002).
Giada L., Marsili M., *Physica A* **315**, 650 (2002).
Bonanno G., Caldarelli G., Lillo F. and Mantegna R.N., Phys. Rev. E **68**, 046130 (2003).
Bonanno G., Caldarelli G., Lillo F., Miccichè S.,Vandewalle N. and Mantegna R.N., *Eur. Phys. J. B* **38**, 363 (2004).
Tumminello M.,Aste T., Di Matteo T. Mantegna, R.N., *Proc. Natl. Acad. Sci. USA* **102**, 10421 (2005).
Coronnello C., Tumminello M., Lillo F., Miccichè S. and Mantegna R.N., Acta Phys. Pol. B **36**, 2653 (2005)
Metha M.L., *Random Matrices* (Academic Press, New York, 1990).
Anderberg M.R., *Cluster Analysis for Applications* (Academic Press, New York, 1973).
Rammal R., Toulouse G., and. Virasoro M.A, Rev. Mod. Phys. **58**, 765 (1986).
Mardia K.V., Kent J.T. and Bibby J.M. *Multivariate Analysis*, (Academic Press, San Diego, CA, 1979).
Gower J.C., Biometrika **53**, 325 (1966).
Papadimitriou C.H. and Steiglitz K., *Combinatorial Optimization* (Prentice-Hall, Englewood Cliffs, 1982).
Mantegna R.N. and Stanley H.E., *An Introduction to Econophysics: Correlations and Complexity in Finance*, (Cambridge University Press, Cambridge, 2000), pp. 120-128.
Gower J.C., Applied Statistics **18**, 54 (1969).
Tumminello M., Lillo F. and Mantegna R.N., manuscript in preparation.
Miccichè S., Bonanno G., Lillo F. and Mantegna R.N., Physica A **324**, 66 (2003).
Di Matteo T., Aste T. and Mantegna R.N., Physica A **339**, 181 (2004).
West D.B., *An Introduction to Graph Theory* (Prentice-Hall, Englewood Cliffs, NJ, 2001).
Dacorogna M.M., Gencay R., Müller U.A., Olsen R.B., Pictet O.V., *An Introduction to High-Frequency Finance*, Academic Press (2001).
Lillo F. and Mantegna R.N., Phys. Rev. E **72**, 016219 (2005).

Chapter 7

Innovation Systems by Nonlinear Networks

P. Andriani, F. Conti, L. Fortuna, M. Frasca, G. Passiante and A. Rizzo

Cellular Neural Networks (CNNs) constitute a powerful paradigm for modelling complex systems. Innovation systems are complex systems in which small and medium (SME) enterprises play the role of simple units interacting each other. In this paper innovation systems based on CNN are investigated. It is shown how a model based on CNN can reproduce the main features of innovation systems and how this model can be generalized to include different aspects of the actors of the financial market.

7.1 Introduction

Complex systems are often defined as systems made of elementary units interacting with simple laws and able to allow the emergence of an holistic global behavior. Recently the theory of complex systems, intended as systems that intrinsically retain the possibility to respond to the environment in different ways, has been applied to innovation systems [Allen (2001)].

The key point is that industrial network have to be adaptive to respond to the complex unpredictable behavior of the market and that innovation plays a fundamental role to achieve adaptability. Survival and success are possible if the elements of the network are able to learn and change.

In this work Cellular Neural Networks (CNNs) [Chua (1993)] [Chua (1988a,b)] modelling innovation systems are introduced. CNNs have been already used to model complex nonlinear phenomena in spatially distributed systems. In fact an n-dimensional CNN may represent any complex system in which interactions among the elements of the system are ruled by local connections and where the main features of the basic unit are reflected by the model assumed for the elementary CNN cell.

The main idea of the approach here presented is to use each layer of a multi-layer CNN to model a particular variable of the financial system. Complexity can be added by adding further layer in the model, i.e. increasing the dimension of the basic cell.

In particular this paper focuses on a very simple model of innovation in systems of interconnected agents. As introduced before this can be extended to account for other characteristics of the elementary agent by adding other CNN layers.

The model for innovation examined in this paper is based on the simple rules discussed in a recent paper by Watts [Watts (2002)]. This model accounts for the presence (and the absence under certain conditions) of global cascades in networks of agents.

Global cascades represent the spread of the innovation along the whole network and are common in many social and economic phenomena. For example it accounts for the fact that some books, movies or songs become very popular despite their initial small marketing budgets [Gladwell (2000)].

This is an example of what often happens in economic and social systems in which the knowledge of the problem or the ability to process the available information are limited and the decision makers have to pay attention to each other [Bikhchandani (1992)].

In the example of a popular hit recommendations of friends and neighbors can play a fundamental role in the choice of a movie, a book or a song. This phenomenon in economics is known as information cascade: individuals make decision on the basis of the action of other individuals of the population, thus allowing the possibility that the whole system exhibits a herd-like behavior.

The model proposed by Watts [Watts (2002)] and here implemented by the CNN paradigm provides a possible explanation for this phenomenon in terms of a binary-state decision random network .

This model is detailed in Section 2. Section 3 deals with the CNN model, while Section 4 presents the results of the simulation of the CNN model, showing how the CNN model can reproduce the main features of the innovation model. Section 5 draws the conclusions of the paper.

7.2 Cellular Automata Model

In this Section the model for innovation proposed by Watts [Watts (2002)] is briefly reviewed. This model extends the model introduced by the sociologist Granovetter [Strogatz (2003)], which illustrated his results on the brink of rioting and assumed that each person's decision is dependent on what everyone else is doing.

This model described a population of agents which must decide between two actions: adopt or not a new technology. Watts used a model based on a cellular automaton (CA)[Wolfram (1986)] to model this system.

Individuals are represented by binary variables that can adopt or not the innovation. If the individual adopt the innovation the cell state of the cellular automata is 0, if don't adopt the innovation the state of the cellular automata state is 1. Connections among individuals are assumed random.

Each individual is characterized by a threshold. In fact, each individual adopts the innovation if the percentage of connected individuals that has already adopted the innovation is greater than its threshold. The threshold is different for each agent and represents the attitude to innovation of the given agent.

This simple model is able to account for the emergence of global cascades, in which innovation spread in the whole network. However global cascades are possible only under particular conditions.

The main result is that the presence of cascades depends on the average number of connections between individuals. When the number of connections is too small there is no room for cascades. When the number of connections increases, the attitude to innovation increases until global cascades become possible. However if the network is very connected global cascades become rare events.

The reason of this behavior is that each individual makes its decision on the basis on a percentage of the individuals with which he is connected and that adopted innovation. Thus increasing the connectivity level two bifurcation points are evident: cascades are therefore possible only if connectivity is neither small or too large.

More in detail some simulation results are here discussed. The number of initial innovators is kept constant (as in Fig. 7.1), and the behavior with respect to different average numbers of connections is investigated.

There is no cascade if the number of connections is too small: in this case innovation does not spread (Fig. 7.2(a) shows an example).

If the connection is increased, cascade propagation becomes possible,

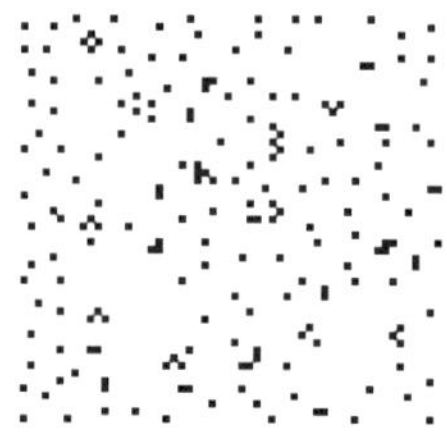

Fig. 7.1 The initial condition of the CA model represents the initial innovators (black pixels indicate that the cell status is 1, thus the individual is an innovator).

innovation does spread in the whole network. Fig. 7.2(b) shows an example of this behavior. If the number of connections is too high, the attitude to innovation of the network decreases, since each agent is connected with many other individuals and so needs many innovators to be convinced. Fig. 7.2(c) is an example of this last case.

(a) (b) (c)

Fig. 7.2 Innovation propagation in a random network with respect to different connectivity levels, the mean number of connections is: 5 (a), 30 (b) and 50 (c). Cascade propagation is possible only in case (b).

The results highlight the presence of two phase transitions when the mean number of connections is increased.

7.3 Innovation Models Based on CNNs

The simple model of cascade propagation is an example of how CNN can be used for modelling complex financial systems.

In this Section we show how CNN can be used to reproduce the main features of the CA model introduced by Watts. Moreover they allow the generalization of the random network to a generic grid of connected units and the extension of the discrete time model to a continuous one.

Different grids can therefore be studied and connections can be assumed local or mostly local leading to the possibility of studying innovation propagation in different topologies.

In the following a regular grid (shown in Fig. 7.3) is taken into account and standard 1-layer templates are used. In fact each innovator can be modelled by the state of a cell of a 1-layer CNN. Now the decision to adopt or not the innovation takes place as a continuous process. The saturation points +1 and -1 represent the decision to adopt the innovation or not, respectively.

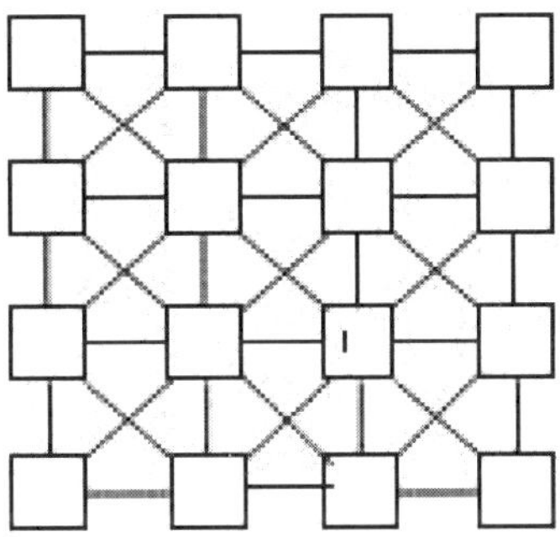

Fig. 7.3 Scheme of a CNN.

The state equation of the cell C(i,j) is the following [Chua (1988a,b)]:

$$\dot{x} = -\frac{x_{ij}}{R_{ij}} + i_1 + \sum(A_{ij:kl}y_{kl} + B_{ij:kl}u_{kl}) \tag{7.1}$$

where:

$$N_r(i,j) = C(k,l) : max(|k-i|, |l-j| < r) \tag{7.2}$$

r is the size of the neighborhood, A and B are the feedback and control template, respectively, and:

$$^7.3i = 1..N, j = 1..M \tag{7.3}$$

To completely specify the model templates have to be designed in order to reproduce the features of the decision maker unit. This step implies a sort of translation between CA rules and CNN templates.

Recently great advances have been done in the direction of an analytical universal method to accomplish this step [Chua (2002)], but until now this step still requires trial and error.

However in this case this step can be accomplished by adapting templates already known. The behavior of each cell depends on the neighbor states in such a way that if the percentage of neighbor innovators is greater than a threshold the cell itself should become an innovator.

This resembles the rules for game of life, and therefore similar templates have been adopted [Crounse (1997)]. In particular the bias value has to be changed to account for the condition on the threshold. In fact it is the bias that mimics the threshold of the individual.

Finally the neighborhood radius r has to be chosen. Since we want to study the effects of different connectivity levels, three different neighborhood radius have been chosen: r=1, r=2 and r=3.

The A and B templates remain qualitatively the same in the three cases, but the bias has to be properly scaled in order to represent a percentage threshold. In particular the bias has been chosen so that it represents a percentage threshold equals to 44 in all the three cases discussed in the following.

In particular, for example in the case of r=1 the templates that we considered are:

$$A = \begin{bmatrix} 0\,0\,0 \\ 0\,1\,0 \\ 0\,0\,0 \end{bmatrix} ; B = \begin{bmatrix} 1\,1\,1 \\ 1\,1\,1 \\ 1\,1\,1 \end{bmatrix} \tag{7.4}$$

and the bias is I=4. For r=3 the A template is a 7x7 matrix with only the central element different from zero, the B template is a 7x7 matrix with all unitary elements and the bias is I=22.

The final output of the CNN model for innovation is the logic OR between the result of the processing of these templates and the input representing the initial state of innovators.

7.4 Simulation Results

As initial distribution of innovators the same initial condition of the CA model represented in Figure 1 has been considered.

The simulation results shown in Fig. 4 confirm that, even in the case of regular grids, cascade propagation does not depend on the number of initial innovators, but on the mean number of the connections.

In particular Fig. 7.4 shows simulation results for a CNN with $r = 1$, $r = 2$ and $r = 3$.

The same scenario depicted by Watts appears, there are two phase transitions: when the connectivity is low ($r = 1$) there are no global cascades; for $r = 2$ global cascades are possible, and for $r = 3$ global cascades occurs rarely.

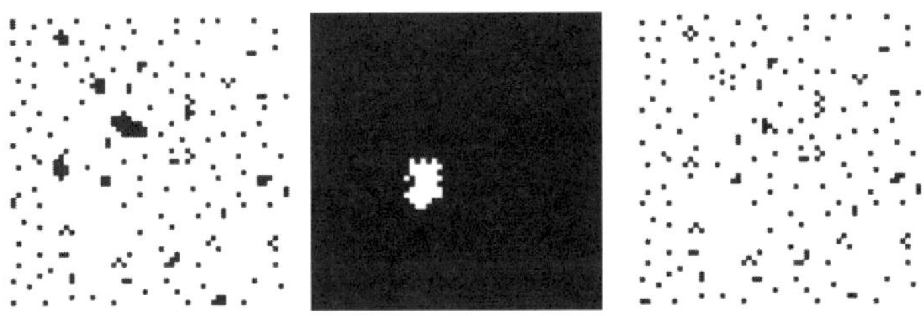

(a) (b) (c)

Fig. 7.4 Simulation result of a CNN with: (a) $r = 1$, (b) $r = 2$ and (c) $r = 3$.

Similar results can be obtained with different initial conditions. This allows us to conclude that cascade propagation does not depend on the number of initial innovators but on individual thresholds and connectivity levels.

7.5 Conclusions

In this paper it is shown how CNNs can be used to model complex system models arising in the field of financial and innovation systems. In particular a suitable CNN implementation of a simple model of innovation [Watts (2002)] has been illustrated.

The use of CNN allows to consider continuous time instead of discrete time models and provides a generalization of the paradigm of connectivity.

In fact by using CNNs different kinds of connections (either regular or almost locally connected systems) can be studied.

Moreover, CNN models can be easily simulated on parallel hardware, thus allowing real-time simulation of complex financial systems.

Another methodological aspect of the CNN approach is the possibility of adding complexity to the basic cell model by adding new CNN layers modeling other interest variables of the elementary agent.

The CNN model for innovation presented as a case study shows the effectiveness of the approach confirming the presence of two phase transitions in the model of innovation.

If the network is poorly connected no global cascade may occur, if the network is too largely connected global cascades become rare events.

References

Allen P., A Complex systems approach to learning in adaptive networks. *Int. J. of Innovation Management*, 5(2):149–180, 2001.

Bikhchandani S., Hirshleifer D. and Welch I., J. Pol. Econ., 100(5): 992–1026, 1992.

Chua L. O., Yoon S. and Dogaru R., A Nonlinear Dynamics Perspective of Wolfram's New Kind of Science Part I: Threshold of Complexity. *Int. J. Bifurcation and Chaos*, 12(12), 2655–2766, 2002.

Chua L.O. and Yang L., IEEE Transactions on Circuits and System I. *Cellular Neural Network: Teory*, 35:1257–1272, 1988a.

Chua L.O., Yang L., and Roska T., *Cellular Neural Network: applications.* IEEE Transactions on Circuits and System I, 35 :1273–1290, 1988b.

Chua, L. O. and Roska T., CNN: The paradigm. *IEEE Transactions on Circuits and Systems I*, 35:147–156, 1993.

Crounse K. R., Fung E. L. and Chua L. O., Efficient implementation of neighbourhood logic for cellular automata via the cellular neural network universal machine. *IEEE Transactions on Circuits and System I*, 44(4), 1997.

Gladwell M., *The Tipping Point: How Little Things Can Make a Big Difference.* Little Brown New York, 2000.

Strogatz S. H., *Sync: The Emerging Science of Spontaneous Order.* Hyperion Press, 2003.

Watts D. J., A simple model of global cascades on random networks. *Proc. Natl. Acad. Sci. 99*, 5766–5771, 2002.

Wolfram S., *Theory and application of cellular automata*, World Scientific, 1986.

Index